THE
CLOUD
BOOK

HOW TO UNDERSTAND THE SKIES

THE
CLOUD
BOOK

HOW TO UNDERSTAND THE SKIES

In association with the
≋ **Met Office**

DAVID & CHARLES

www.davidandcharles.com

RICHARD HAMBLYN

Contents

FOREWORD

"When the clouds come from the north, there's a chance it might snow", my Dad once said to me as a snow-obsessed child. I must have been about four or five years old, but I soon learned which parts of the sky to be watching, and developed the inevitable understanding that reality was much more complex than that. He often seemed to be right, though!

Humans have always looked to the skies to try and gain some advance warning of what nature may be about to toss up next. From the foreboding dark base and spectacular Cirrus anvil of an approaching Cumulonimbus cloud to the first chink of blue sky spotted through a layer of Stratus as it lifts and breaks on a spring morning, we often get some sense of what might be about to unfold.

Clouds are visible signatures of many kinds of physical processes taking place in the atmosphere as warmth and moisture are constantly redistributed – part of nature's will to even out differences. You may have noticed, on a sunny spring or summer's day, the first puff of Cumulus cloud 'bubble up' at, say, 10 o'clock. On one day, many such clouds might develop quickly, spreading out into a layer of Stratocumulus by noon and blotting out the Sun. The next day, this may not occur – blue skies prevail. Have you ever wondered why? Or, if you live near the coast, have you ever noticed that in summertime such convective clouds form preferentially inland, leaving the coast clear?

Some clouds are made of liquid water; others ice, or a mixture of the two. Some produce rain, snow or hail.

Many produce absolutely nothing, despite their dark appearance. You may have developed a sense of this without ever explicitly wondering why. With spectacular photography and insightful explanations, this book will help the reader connect the scientific understanding with the beauty.

Paul Gundersen
Chief Operational Meteorologist
Met Office

INTRODUCTION

Clouds and their Classification

Clouds have been objects of delight and fascination throughout human history, their fleeting magnificence and endless variability providing food for thought for scientists and daydreamers alike. 'The patron goddesses of idle men', as the playwrite Aristophanes described them in 420 BCE, clouds and the ever-changing patterns they create have long stood as potent symbols of the restlessness and grandeur of nature.

But in contrast to all other earthly phenomena, from microbes and minerals to the greatest plants and animals, every known species of which had been classified and reclassified many times over since early antiquity, clouds (at least in Western culture) remained uncatalogued and unnamed until the early nineteenth century, when the Latin terms that are now in international use — 'Cirrus', 'Stratus', 'Cumulus', and their compounds — were bestowed on them by Luke Howard (1772–1864), an amateur meteorologist from East London.

Luke Howard was not, of course, the first to attempt to understand clouds in a systematic way. Scientific thinkers had long sought to explain the complex mechanics of cloud formation — Aristotle, for example, came up with the theory of atmospheric exhalations, based on the four stratified elements of earth, air, fire and water, with their associated interactive properties of heat and cold, dryness and moisture — but no one had ever hazarded a system of classifying or naming their apparently limitless varieties. This must have been due, at least in part, to the challenge posed by their fleeting instability. Clouds change their form and structure, minute by minute, their shapes appearing 'as indistinct as water is in water', as Shakespeare described them in *Antony and Cleopatra*; so how could objects which remain in a state of constant flux and flow ever be granted permanent or meaningful identities?

The problem was solved in 1802, when the thirty-year-old Quaker Luke Howard (who was a pharmacist by profession, but a meteorologist by inclination) devised a deceptively simple classificatory system, which overcame the challenge of the clouds' continual merging and demerging, as they rise, fall and spread through the atmosphere, rarely maintaining the same shape for more than a few minutes at a time. In contrast to earlier natural history taxonomies, in which genera and species were arranged in fixed relationships, Howard's new classification needed to allow for all this continual movement and change, since, as he expressed it at the time, 'the same aggregate which has been formed in one modification, upon a change in the attendant circumstances, may pass into another.' [1]

Clouds, he noted, change their shape according to invisible processes going on in the atmosphere, an observation which, on its own, would not have been enough to furnish a new classification, but it was accompanied by another, equally penetrating insight that clouds might have many individual shapes, but only a very few basic forms. In fact, claimed Howard, all clouds belong to one of only three principal families, to which he gave the Latin names: *Cirrus* (a word meaning 'fibre' or 'hair'), *Cumulus* ('heap' or 'pile'), and *Stratus* ('layer' or 'sheet'). Every kind of cloud, he claimed, is either a modification of or a transition between one or more of these three major types, with the intermediate or transitional forms named according to their relation to the principal clouds. So a high, wispy Cirrus cloud that descended and spread into a sheet (or stratum) was named Cirrostratus, while groups of fluffy Cumulus clouds that joined up and spread across the sky (again, into a stratum or sheet) were named Cumulostratus. So far, so brilliantly simple. The clouds might still be never at rest, but Howard had arrived at an elegant solution to the problem of naming transitional forms in nature.

Once published, in 1803, Howard's cloud classification was soon taken up across the scientific world, from where it spread into the wider culture through lectures, newspapers and printed books. By the early 1820s the landscape painter John Constable was using his own, heavily annotated copy of Howard's classification in connection with the sequence of more than 100 cloud studies that he painted in the open air at Hampstead Heath during summers he spent in London. Constable took his meteorological studies seriously, claiming that 'we see nothing truly till we understand it', and his great 'six-footer' canvases, on which he staked his reputation as a painter, became increasingly dominated by vast, eventful skies, the precise dispositions of which had been carefully researched. 'You can never be nubilous', as he once claimed in a letter to a friend, 'for I am the man of clouds'. [2]

The German poet and natural philosopher JW von Goethe apparently felt much the same way, and wrote fan letters to Howard during the early 1820s, as well as a celebratory cloud-poem, *Howards Ehrengedächtnis* ('In Honour of Howard'), which invoked each of the cloud-types in turn, a device that Percy Shelley also deployed in his 84-line poem 'The Cloud' (1820), which describes the varied and distinctive personalities of each of Howard's cloud types, from low Stratus:

From my wings are shaken the dews that awaken
The sweet buds every one

to high Cirrocumulus:

When I widen the vent in my wind-built tent
Till the calm rivers, lakes and seas,
Like strips of the sky fallen through me on high,
Are each paved with the Moon and these

via each of the others, including my favourite, Cumulonimbus:

I wield the flail of the lashing hail,
And whiten the green plains under,
And then again I dissolve it in rain,
And laugh as I pass in thunder. [3]

OPPOSITE:
John Constable,
Study of Cirrus Clouds,
oil on paper, c.1822.
Constable owned a copy of Luke Howard's cloud classification, and on the reverse of this oil sketch, painted in the open air on Hampstead Heath, he has written the word 'Cirrus'.

Classification of Cloud Forms from the First
International Cloud Atlas, 1896

a. Separate or globular masses (most frequently
    seen in dry weather)
b. Forms which are widely extended, or
    completely cover the sky (in wet weather)

**A. Upper Clouds, average altitude 9,000m
(29,500ft)**

a. 1. Cirrus
b. 2. Cirro-stratus

**B. Intermediate Clouds, between 3,000
and 7,000m (9,840 and 23,000ft)**

a. 3. Cirro-cumulus
a. 4. Alto-cumulus
b. 5. Alto-stratus

C. Lower Clouds, 2,000m (6,500ft)

a. 6. Strato-cumulus
b. 7. Nimbus

D. Clouds of Diurnal Ascending Currents

8. Cumulus; apex, 1,800m (5,900ft);
base, 1,400m (4,600ft)
9. Cumulo-nimbus; apex, 3,000 to 8,000m
(9,840 to 26,240ft); base, 1,400m (4,600ft)

E. High Fogs, under 1,000m (3,280ft)

10. Stratus

Meteorology itself, meanwhile, was beginning to be organized on an international footing, and as the nineteenth century wore on, Howard's original classification was gradually refined and enlarged in accordance with new insights and observations into the behaviour of clouds and weather. The first of many changes to be made was the addition of Stratocumulus, a term suggested in 1840 by the German meteorologist Ludwig Kaemtz, who was keen to distinguish those rolling masses of greyish cloud from what Howard had termed 'Cumulo-stratus: the Cirro-stratus blended with the Cumulus'. Kaemtz's inversion of the compound terms removed the cloud from the convective Cumulus family and placed it in the category of Stratus, assigning it to a more suitable position within the family of low pressure clouds. Subsequently, and by general agreement, Howard's original term ('Cumulo-stratus') was dropped from the list in favour of Stratocumulus, now defined as 'a layer of cloud, not flat enough to be called pure Stratus, but rising into lumps too irregular and not sufficiently rocky to be called true Cumulus'.

Not long after, in 1855, Émilien Renou, director of the French observatories at Parc Saint-Maur and Montsouris, made two further additions to the growing classification in the form of Altocumulus and Altostratus, the term *alto* deriving from the Latin word for 'elevated'. Both these new cloud genera, as Renou pointed out, were medium-level clouds, with their altitude emphasized in the names he bestowed, since, as he argued, their altitude had such a crucial shaping influence on their form. Renou's suggestion gave strength to the case for adopting altitude as the principal criterion for grading the families of cloud. This idea was soon taken up by observers all over the world, and in September 1896 — 'The International Year of Clouds' — the International Meteorological Congress (IMC), at their annual conference in Paris, formally adopted an expanded version of Howard's

OPPOSITE:
**Engravings from Luke Howard's
*Essay on the Modifications of
Clouds* (1804).** The first (left)
shows the main cloud types:
Cirrus, Cumulus and Stratus.
The second (right) shows the
compound forms: Cirrocumulus,
Cirrostratus and Cumulostratus.

13

ABOVE:
Portrait of Luke Howard – circa 1807, attributed to John Opie, one of the leading society portraitists of the day.

original seven-part classification as the official global standard, in which altitude was promoted as the primary basis of cloud identification. The new classification was outlined in the first *International Cloud Atlas*, a multi-lingual guidebook published by the IMC soon after the Paris conference, which featured the following five-part altitude scheme, designed not only to distinguish between detached and continuous cloud formations ('separate or globular masses' versus 'forms which are widely extended', as they expressed it), but also to accommodate clouds which tended to rise or fall between the height bands, as is often the case with convective cumuliform clouds.

This was how the expression 'to be on cloud 9' was introduced to the language: to imagine being on top of a Cumulonimbus, the highest-climbing cloud of all (ranked number 9 in the newly published classification), really was to imagine being 'on top of the world'. Subsequent editions of the *Cloud Atlas*, however, saw Cumulonimbus renumbered as cloud 10 (with the cloud 9 spot taken over by Cumulus, a generally low-lying convective cloud), although happily, the World Meteorological Organization (WMO) has since relented, and renumbered the ten cloud genera from 0–9, so the top of cloud 9 remains, meteorologically as well as figuratively, the highest (and thus the happiest) spot in the sky.

Current World Meteorological Organization Classification of the Ten Principal Cloud Genera

High Clouds, base usually above 6km/>20,000ft
0. Cirrus (Ci)
1. Cirrocumulus (Cc)
2. Cirrostratus (Cs)

Medium Clouds, base usually between 2 and 6km/6,500 and 20,000ft
3. Altocumulus (Ac)
4. Altostratus (As)
5. Nimbostratus (Ns)

Low Clouds, base usually below 2km/<6,500ft
6. Stratocumulus (Sc)
7. Stratus (St)
8. Cumulus (Cu)
9. Cumulonimbus (Cb)

OPPOSITE:
Engraving from Luke Howard's *Essay on the Modifications of Clouds* (1803), showing a Cumulonimbus raincloud passing over a highly romanticized landscape. Howard originally termed this cloud 'nimbus', but it was renamed Cumulonimbus later in the nineteenth century.

15

In addition to the temporary confusion over the true identity of cloud 9, a number of other significant changes have also been made to the official classification, particularly when it comes to the increasing refinement of our understanding of cloud species and varieties. So far we have mentioned only the ten cloud genera — the ten principal types (see page 14) — but part of the ongoing success of Howard's classificatory scheme is that it has proved compatible with the international system of binomial nomenclature that was introduced by the Swedish botanist Carl von Linné (known to the Latinizing world as Linnaeus) in the early eighteenth century.

Linnean taxonomy is based on the organizing concept of a genus (plural 'genera': a class or group of organisms with common characteristics) that is then subdivided into two or more species. For example, the generic word 'heron' (genus: *Ardea*) can be applied to a range of similar-looking birds, each one of which will be a member of a particular species, such as the Grey heron (*Ardea cinerea*) or the Great blue heron (*Ardea herodias*). Latin names such as *Ardea cinerea* are thus made up of the name of the genus followed by the name of the species, which often contains a descriptive clue (*cinerea* derives from the Latin word for 'embers'). On the same principle, most clouds can be identified as belonging to a particular

species as well as to one of the ten main genera, with its specific name alluding to a characteristic shape or structure. A Cirrus uncinus, for example, is an easily recognised hook-shaped species of Cirrus cloud (see C$_H$1) (*uncinus* deriving from the Latin word for 'hooked'), while a Cirrostratus nebulosus is a misty or nebulous form of Cirrostratus (see C$_H$7) (*nebulosus* being the Latin word for 'misty' or 'indistinct').

Cloud *varieties*, meanwhile, are used to identify certain additional characteristics, such as a cloud's relative transparency, or a particular arrangement of its elements. Altocumulus stratiformis duplicatus, for example (see C$_M$7) refers to a stratiform species of Altocumulus that occurs in two or more layers (*duplicatus* deriving from the Latin word for 'doubled'), while a single layer of the same cloud that was thin enough to allow sunlight to pass through it would be given the varietal term *translucidus*. The following chart shows the fifteen species and nine varieties most often associated with the ten cloud genera, while the glossary beginning on page 166 explains the meaning and derivation of every term used in the cloud vocabulary. At first glance, all this Latin might look hard to take on board, but it is in fact remarkably straightforward, since most of the equivalent terms in English share the same Latin root, such as *floccus* (meaning 'tufted' or 'flocked'); *congestus* ('growing

upwards'); or *duplicatus* ('in more than one layer'). It certainly need not be an obstacle to those who (like me) missed out on a classical education. And as Luke Howard himself pointed out in 1818, in the preface to the first volume of *The Climate of London*, 'the names for the clouds which I deduced from the Latin are very easy to remember, and the meaning of each was carefully fixed by a definition: the observer having once made himself master of this, was able to apply the term with correctness, after a little experience, to the subject under all its varieties of form, colour, or position'. [4]

The Cloud Species and Varieties Associated with the Ten Cloud Genera

Low Clouds

Cumulus (Cu)
humilis (C_L1)
fractus (C_L1)
congestus (C_L2)
mediocris (C_L2)
radiatus
flammagenitus
homogenitus

Cumulonimbus (Cb)
calvus (C_L3)
capillatus (C_L9)

Stratocumulus (Sc)
cumulogenitus (C_L4)
stratiformis (C_L5)
castellanus (C_L5)
lenticularis (C_L5)
floccus
volutus
translucidus
perlucidus
opacus
duplicatus
undulatus
radiatus
lacunosus

Stratus (St)
nebulosus (C_L6)
fractus (C_L7)
opacus
translucidus
undulatus
cataractagenitus

Medium Clouds

Altostratus (As)
translucidus (C_M1)
opacus (Nimbostratus) (C_M2)
duplicatus
undulatus
radiatus

Nimbostratus (Ns) (C_M2)

Altocumulus (Ac)
lenticularis (C_M4)
stratiformis (C_M5)
castellanus (C_M8)
floccus (C_M8)
translucidus (C_M3)
duplicatus (C_M7)
volutus
perlucidus
opacus
undulatus
radiatus
lacunosus

High Clouds

Cirrus (Ci)
uncinus (C_H1)
fibratus (C_H1)
floccus (C_H2)
castellanus (C_H2)
spissatus (C_H3)
intortus
radiatus
vertebratus
duplicatus
homogenitus

Cirrostratus (Cs)
nebulosus (C_H5, C_H6, C_H7, C_H8)
fibratus (C_H5, C_H6, C_H7, C_H8)
duplicatus
undulatus

Cirrocumulus (Cc)
stratiformis (C_H9)
lenticularis (C_H9)
castellanus (C_H9)
floccus (C_H9)
undulatus
lacunosus

How to use this Book

This book follows the internationally recognized cloud coding convention that was introduced by the World Meteorological Organization (WMO) in the 1939 edition of the *International Cloud Atlas*. It assigns a unique identification code (and accompanying symbol) to each cloud species or modification, with the dropped middle letter denoting the altitude band to which the cloud belongs. Thus, Low Clouds (bases usually below 2km/<6,500ft) are coded from C_L1 to C_L9; Medium Clouds (bases usually between 2 and 6km/6,500 and 20,000ft) are coded from C_M1 to C_M9; and High Clouds (bases usually above 6km/>20,000ft) from C_H1 to C_H9.

The principal advantage of employing this coded classificatory scheme, rather than the basic ten-genera system, is that it places a much greater emphasis on the processes of cloud growth and decay that so determine the ever-changing appearance of our skies. As will be seen throughout Part I, many of these individual classifications describe changing states of the sky as a whole, often tracing the eventful careers of certain cloud species over extended periods of time. This is particularly the case within the complex arena of the high- and medium-level clouds. Slow-moving Cirrostratus, for example, occupies four of the 27 coding bands, ranging from C_H5

('Cirrostratus increasing but below 45° elevation') to C_H8 ('Cirrostratus neither progressively invading the sky nor entirely covering it'), while the many modifications of Altocumulus occupy no fewer than seven separate coding bands, from C_M3 ('Altocumulus translucidus and Altocumulus stratiformis at a single level') to C_M9 ('Altocumulus of a chaotic sky'), chaotic being a recognized meteorological term denoting a fragmented mixture of poorly defined transitional clouds occurring at several levels. As the *Cloud Atlas* observed of its new cloud code in 1939, 'it would perhaps be more appropriate to call it a code of the types of skies', although, of course, 'what really characterizes the type of sky is the aggregate of individual clouds and their *organization*'.

Understanding such organization is the key to understanding cloud behaviour, and in following the WMO's coding convention, this book is intended to enable the reader not only to identify individual clouds and skies as they might appear at any given moment, but also to track their likely changes over time. For example, the characteristic high, icy clouds of the category C_H1 – Cirrus uncinus and Cirrus fibratus (see page 82) – might well decay and disappear by slowly sinking into warmer air and subsequently evaporating, but sometimes, when a

depression is on its way, and a layer of moist air is rising high, these clouds can spread for miles across the sky, transforming into clouds of the category C_H4 (see page 92). These are essentially the same clouds as C_H1, but the character of the sky has become entirely different, and thus deserves a separate definition. Cloud behaviour – their capacity to transfer allegiance from one modification to another within the space of a few minutes or hours – is an integral part of cloud classification, a discipline which, in common with all other branches of meteorology, puts great emphasis on the careful observation of processes unfolding over time. For when it comes to clouds and weather, there is never a moment when nothing can be said to be happening.

In addition to the ten genera, fifteen species and nine varieties of cloud that populate Part I of this book, there are also four accessory clouds: pileus ('cap cloud'), pannus ('scud'), velum ('veil cloud'), and flumen ('flow cloud') (see pages 110–113), which occur only in conjunction with the principal types, as well as eleven supplementary features: arcus ('arch'), asperitas ('roughened cloud'), cauda ('tail'), cavum ('fallstreak hole'), fluctus ('wave cloud'), incus ('anvil'), mamma ('udders'), murus ('wall cloud'), praecipitatio ('rain'), tuba ('funnel') and virga ('fallstreaks') (see pages 114–127), most of which are occasional players in the riotous displays put on by large Cumulonimbus clouds. All of these are described in full in Part 2, as are a number of special clouds such as the mysterious noctilucent cloud (see page 130), or the man-made variety of high Cirrus clouds created by aircraft contrails (see page 138), as well as, for good measure, a series of optical effects and phenomena that are associated with cloud activity, such as haloes, parhelions, and crepuscular rays. The fact that twelve of these cloud names were only added to the official classification in 2017 shows that clouds remain a vital subject for scientific research, not least when considered within the wider context of global environmental change. The Afterword on pages 155–164 discusses the role of clouds in climate change, an area of research that has become increasingly urgent in recent years, and that has served to position clouds at the heart of global atmospheric inquiry.

In short, all things nephological — all things to do with the life of a cloud, whether a giant Cumulonimbus or a tiny shred of Stratus fractus, whether an everyday occurrence or a fleeting rarity, and whether fully understood or entirely baffling — are to be found in the following pages. 'For if there is nothing new on Earth', as the pioneer ecologist

Henry David Thoreau noted in his nature journal on 17 November 1837, 'still there is something new in the heavens. We have always a resource in the skies. They are constantly turning a new page to view. The wind sets the types in this blue ground, and the inquiring mind may always read a new truth.' [5]

---

[1] Luke Howard, *On the Modifications of Clouds &c.*, (London, 1804): p. 4

[2] Cited in E. Morris (ed.), *Constable's Clouds: Paintings and Cloud Studies by John Constable* (Edinburgh, 2000): p. 169

[3] See John E. Thornes, *John Constable's Skies: A Fusion of Art and Science* (Birmingham, 1999): p. 190–191

[4] Luke Howard, *The Climate of London* (London, 1818): p. xxxii

[5] *The Journal of Henry D. Thoreau*, ed. B. Torrey & F. H. Allen, 14 vols (New York, 1962): I, p.11

CLOUD CLASSIFICATION SYMBOLS

Each of the 27 cloud states has been assigned a unique international symbol, which is used by meteorologists and aviators as a form of rapid visual shorthand when reporting the state of the sky.

LOW	MEDIUM	HIGH
C_L1	C_M1	C_H1
C_L2	C_M2	C_H2
C_L3	C_M3	C_H3
C_L4	C_M4	C_H4
C_L5	C_M5	C_H5
C_L6	C_M6	C_H6
C_L7	C_M7	C_H7
C_L8	C_M8	C_H8
C_L9	C_M9	C_H9

PART 1:
THE PRINCIPAL
CLOUDS

Low Clouds

(of genera Stratocumulus, Stratus, Cumulus and Cumulonimbus)

C_L 1

> **SPECIFICATION:** Cumulus clouds with little vertical extent and seemingly flattened (humilis), or ragged Cumulus other than that of bad weather (fractus).
>
> Symbol = ⌒

Cumulus fractus; Cumulus humilis ('fair weather clouds')

One of my favourite *Peanuts* cartoons features Linus and Charlie Brown, lying on their backs, gazing up at the passing clouds. When Charlie asks Linus if he can see any shapes in them, Linus replies that he's just spotted the outline of British Honduras, the profile of the painter Thomas Eakins, and a remarkably detailed tableau of the stoning of St Stephen: 'there's the Apostle Paul standing to one side. What about you, Charlie Brown?': 'I was going to say I saw a ducky and a horsie, but I just changed my mind.'

RIGHT:
This photograph, taken on the beach at Portreath, Cornwall, on a summer afternoon, shows a group of **Cumulus fractus** (Cu fra) clouds at an early stage of their formation, rising on thermals from the Sun-warmed sea.

OPPOSITE:
A quiet summer sky filled with **Cumulus humilis** (Cu hum) clouds, passing over the Vale of Evesham, Worcestershire.

The source of Charlie Brown's mortification was a fleet of small Cumulus clouds glowing in the Sun, the largest of which, as can be inferred from Linus's martydom tableau, had already graduated into larger Cumulus congestus clouds (as described in the following entry). These convective clouds are formed above thermals (columns of ascending air) that rise in plumes from the Sun-warmed ground, with the smaller Cumulus fractus clouds sometimes seen emerging from fragments of haze on warm, calm summer mornings. Seeding themselves on condensation nuclei – microscopic grains of dust, smoke, pollen or sea-salt that are naturally present in the air – the rising pockets of water vapour cool and condense into droplets, which then begin to coalesce, growing upwards and outwards into a puffy white cloud. When completely formed, these dense, white, detached clouds, with wide areas of blue sky between them, have clear-cut horizontal bases and rounded tops, and are often referred to as 'fair-weather' Cumulus, particularly if they rise and consolidate to become Cumulus humilis clouds, as in the photograph on page 23, the bases of these distinct rows of Cumulus humilis clouds hanging some 600m (2,000ft) above the ground.

Over land, on clear mornings, cloudlets of Cumulus may form as the Sun rapidly heats the ground, or may result from the transformation of patches of foggy Stratus nebulosus (see C_L6). If these smaller Cumulus formations begin to show moderate vertical development, especially on warmer afternoons after rain, when the atmosphere may be growing a little unstable, they begin to be classified as Cumulus mediocris or Cumulus congestus (see C_L2), the main difference between this trio of species being that Cumulus humilis clouds are wider than they are tall, while Cumulus mediocris clouds are as tall as they are wide, and Cumulus congestus clouds are even taller. But not all C_L1 clouds are destined to grow into C_L2 species, and at the end of calm summer days when the warm air begins to cool at sunset, and the thermals cease their rising, both these Cumulus species will begin to sink and dissipate, breaking down into ever smaller fragments. Since neither of these smaller Cumulus species are rain-bearing clouds, their evening disappearing act is due to changes in air temperature, not to any shrinkage through precipitation.

OPPOSITE:
Cumulus fractus
clouds beginning to dissipate at
the end of a summer evening.

C$_L$2

> **SPECIFICATION:** Cumulus clouds of moderate or strong vertical extent, generally with protuberances in the form of domes or towers, sometimes accompanied by other cumuliform clouds, all with their bases at the same level.
>
> Symbol =

Cumulus congestus; Cumulus mediocris

LEFT:
Puffy white **Cumulus congestus** (Cu con) clouds grow rapidly through upward convection on sunlit days.

OPPOSITE:
Cumulus mediocris (Cu med) clouds, arranged by the wind into parallel lines known as 'cloud streets', Dover harbour, Kent.

Following on from the previous entry, these larger Cumulus clouds are developments of C$_L$1 clouds, formed by the upward convection of columns of warm, moist air on sunlit days. As these thermals rise, they expand and cool until they reach the dew point (the temperature at which water vapour condenses into droplets), at which point their payload of moisture condenses and coalesces into clouds. The condensation process releases a great deal of latent heat (thermal energy locked up in water vapour), which serves to warm the air inside the growing cloud, leading to further buoyant convection and thus to further build-up of the cloud. The more this cycle continues, the taller the cloud will grow, especially if the process begins early in the morning, with a full day of sunshine ahead of it. The cumuliform cycle depends to a great extent on the

stability and temperature of the surrounding air: if the rising moisture matches the temperature of warm, stable surrounding air, it will tend to spread out, into stratiform clouds, rather than grow upwards; but if the rising moisture remains surrounded by cooler air, it will carry on rising to form cumuliform clouds, a sure sign of atmospheric instability. There are plenty of external limits to Cumulus cloud growth, however, such as wind shear from the side, evaporation from above, and uneven convection from below (varied surfaces on the ground reflect varying amounts of heat radiation), so that some Cumulus clouds will grow only to the size seen in the photograph on page 26, or the Cumulus mediocris clouds that have been arranged by the wind into parallel lines known as 'cloud streets' (or Cumulus mediocris radiatus, to give them their full Linnean designation).

With a day's worth of uplift under their belts, however, energetic Cumulus clouds can grow extremely large, as can be seen in the photograph opposite, their great white convective turrets rearing a mile or more into the sky. Over land, such clouds may well have begun to disperse by the early evening, as thermal convection currents rapidly diminish, but over the oceans, as in this case, Cumulus growth often carries on late into the night, as the sea gives up its absorbed heat radiation over much longer periods. If formed at sea, Cumulus clouds of this size (more than 2km/6,500ft high), can go on to produce light showers, and sometimes even heavy rainfall in the tropics, with larger Cumulus congestus often graduating either upwards into Cumulonimbus calvus clouds (see following entry, C_L3), or outwards, due to horizontal spreading at a temperature inversion, to form thick layers of Stratocumulus cumulogenitus (see C_L4).

RIGHT:
A vigorous **Cumulus congestus** formation massing over Colorado Springs, Colorado.

OPPOSITE:
Ranks of **Cumulus congestus** clouds towering like giants over the sea off Lincolnshire, their turrets rearing a mile or more into the sky.

C$_L$3

SPECIFICATION: Cumulonimbus cloud, its summit lacking sharp outlines, being neither clearly fibrous, nor in the shape of an anvil.

Symbol =

Cumulonimbus calvus

Vast, roiling structures whose summits approach the upper limits of the troposphere, Cumulonimbus calvus clouds represent the next stage of development of large Cumulus congestus clouds. As the congestus clouds continue to build, their tops begin to lose the clearly defined 'cauliflower' appearance, becoming smoother and more consolidated, which, for cloud identification purposes, is the moment when a Cumulus congestus (C$_L$2) classification gives way to Cumulonimbus calvus (C$_L$3), *calvus* deriving from the Latin word for 'bald.' This moment of transition from a C$_L$2 to a C$_L$3 is captured in the photograph of an energetic young Cumulonimbus calvus on the right. It is clearly still growing fast, as indicated by the vigorous protuberances, and the blue sky behind shows that there is still plenty of solar power at its disposal.

RIGHT:
The moment of transition from a Cumulus congestus to a **Cumulonimbus calvus** (Cb cal) is captured here, as a young Cumulonimbus cloud rears over the Ingram Valley, Northumberland.

These clouds can grow to considerable heights, as can be seen in this second photograph, of a much larger Cumulonimbus calvus massing over North Kansas, USA, the summit of which still appears to be swirling with convective energy, although it is also beginning to spread outwards, suggesting that it is in yet another transitional phase, and that this cloud is rapidly on the way to developing into a full-blown stormcloud, complete with an icy storm-brewing thunderhead that will form from the freezing of the summit into clearly visible wind-blown striations (see Cumulonimbus capillatus, C_L9).

From a distance it is easy enough to distinguish a Cumulonimbus calvus from its stormier relation, the Cumulonimbus capillatus, although it is another matter if one happens to be immediately beneath it. Cb cal clouds can produce heavy rain and squalls, but they rarely produce lightning or hail, which tend to be the unique province of Cb cap clouds. But, given that calvus clouds can on rare occasions produce lightning and hail, it can sometimes be impossible to distinguish clearly between a C_L3 and a C_L9; if this is the case, the coding, by convention, is recorded as $C_L=9$.

Both species of Cumulonimbus cloud tend to be given a wide berth by aircraft, since strong currents within them create powerful turbulence, while their high water content can also result in thick layers of ice forming on the cold metal.

BELOW:
A **Cumulonimbus calvus** cloud in a later stage of development, its summit visibly swirling with vigorous convective energy.

C$_L$4

> **SPECIFICATION:** Stratocumulus formed from the spreading out of Cumulus clouds, the remains of which may also be apparent in the sky.
>
> Symbol = ⌒

Stratocumulus cumulogenitus

OPPOSITE:
Stratocumulus cumulogenitus (Sc cugen) is formed when rising Cumulus congestus clouds begin to spread out horizontally, as can be seen in this image of a late afternoon sky over Bracknell, Berkshire.

LEFT AND OVERLEAF:
As the warmth of the Sun decreases at evening, Cumulus clouds often flatten into patches of **Stratocumulus**, a process that can be seen in this sequence of four images taken over a 20-minute period at Totland, Isle of Wight. Patches of Cirrus and Cirrostratus clouds also appear in these photographs.

A prime example of a cloud in transition, as it changes from one genera to another, this species of Stratocumulus occurs when the upper parts of rising Cumulus congestus clouds (see C$_L$2) encounter what is known as a temperature inversion – a layer of air in which the normal rate of atmospheric cooling has either slowed down or gone into reverse, due to the presence of warm air currents moving at higher altitudes. On the whole, rising thermals are not strong enough to punch their way through a temperature inversion, although sometimes some isolated patches of Cumulus growth do resume above the flat-topped layer of Stratocumulus (see also the sky state described in C$_L$8). Upon meeting this barrier to upward convection (air is a very poor conductor of heat), the rising cloud begins to spread horizontally instead, creating a characteristic tapered appearance.

After a while, the separate bases of the Cumulus clouds link up, creating an extensive cloud structure that can cover a large area of sky.

C$_L$5

SPECIFICATION: Stratocumulus not resulting from the spreading out of Cumulus clouds.

Symbol =

Stratocumulus stratiformis; Stratocumulus castellanus; Stratocumulus lenticularis; Stratocumulus volutus

Stratocumulus clouds are the most common clouds on Earth, being routinely visible over vast tracts of land and sea. Originally classified as 'Cumulo-stratus' by Luke Howard in 1803, the Stratocumulus designation was brought in during the mid-nineteenth century in order to emphasize the layered nature of these widespread transitional formations.

In contrast to Stratocumulus formed from the spreading out of Cumulus clouds in the previous entry, these species tend to form from the lifting or breaking up of low sheets of Stratus (C$_L$6) by upwards convection, which thicken into dark rolls, their edges merging to form an apparently continuous layer of cloud; this can be seen in the first two photographs, of a sheet of Stratocumulus stratiformis

TOP:
Stratocumulus stratiformis
(Sc str) form when layers of low-lying Stratus rise and thicken into dark, rolling layers of cloud. Oslofjord, Norway.

RIGHT:
Stratocumulus stratiformis, base 760m (2,500ft), in an evening sky over Anglesey, North Wales.

OPPOSITE:
Patches of blue sky visible between the **Stratocumulus** cloudlets show where pockets of cold air are beginning to sink. Holt, Norfolk.

looming over Oslofjord, Norway, and of a lowering sunset over Anglesey, in North Wales. These layer clouds look as if they might be threatening heavy rain, but in fact they would need to grow much thicker and darker in order to produce anything but the lightest scattering of drizzle.

The same cloud species can also appear in a less continuous form, with blue sky visible between the elements, where pockets of cool air are sinking, as in the photograph on page 37 taken at Holt, in Norfolk, of a sky filled with ragged white clumps of Stratocumulus stratiformis. This formation bears a slight resemblance to Altocumulus stratiformis (see C_M5), the cloudlets of which, however, are smaller and higher, as well as more clearly defined than these, so the two species ought to be easily distinguishable.

Stratocumulus castellanus features convective cumuliform turrets rising from a horizontal base, as can be seen in the photograph on the right. If convection continues, these turrets may grow upwards over the course of a day, developing into Cumulus congestus (C_L2), or even Cumulonimbus clouds, and from there, perhaps, into Stratocumulus cumulogenitus, thus coming full circle from Sc to Cu and back to Sc. As has been said before, clouds are easily capable of changing their allegiance from one genus or species to another, and then back again an hour or two later, especially if forming in unstable air. It's one of the things that makes cloud observing such an endlessly rewarding occupation.

TOP:
Stratocumulus lenticularis (Sc len) clouds form in layers of air rising over hills, as in this example of a cloud-filled sky above St Andrews, Fife.

RIGHT:
Turrets of **Stratocumulus castellanus** (Sc cas) rise from a horizontal base, sometimes growing upwards to develop into Cumulus congestus clouds.

OPPOSITE:
Stratocumulus volutus ('roll cloud') off the coast of Punta del Este, Uruguay.

Stratocumulus lenticularis is the least common species in the Sc genus. It forms in moist air that rises gently over hills, creating long, smooth, lenticular shapes, as can be seen in the first photograph (see page 38). This kind of wave cloud looks very different from Altocumulus lenticularis (see C$_M$4), being more undulating, and extended in length, and thus not so UFO-like.

But the rarest species of all the Stratocumulus clouds is the so-called 'roll cloud', Stratocumulus volutus, which is closely related to the supplementary feature known as arcus, or shelf cloud (see page 117). Unlike arcus, which is always attached to its parent storm cloud, volutus is free-floating, taking the form of a long, detached, tube-shaped cloud mass that often appears to roll slowly about a horizontal axis. Sc vol usually occur singularly but have occasionally been observed in parallel formations.

C$_L$6

> **SPECIFICATION:** Stratus in a more or less continuous sheet or layer, or in ragged sheets, or both, but no Stratus fractus of bad weather.
>
> Symbol = ———

Stratus nebulosus

Stratus nebulosus is a low-level cloud formation made up entirely of water droplets, that appears as a uniformly grey, featureless layer that can extend across the sky for many miles. In contrast to convective cumuliform clouds, low Stratus clouds form in cool, stable conditions, with no pockets of turbulent activity going on. Instead they tend to form when gently rising breezes carry cool, moisture-laden air across a cold sea or land surface, causing widespread, low-level condensation to occur, generally well below c. 500m (1,640ft). As can be seen in the photograph (opposite) of Boston, Massachusetts, it is often low enough to obscure the tops of trees and buildings. Although these clouds can sometimes form from an overnight ground fog lifted

RIGHT:
Moonlight shining through a thin veil of **Stratus nebulosus translucidus** (St neb tr).

OPPOSITE:
A layer of **Stratus nebulosus** (St neb) draped over Boston, Massachusetts, obscuring the tops of buildings, and no doubt making downtown residents feel decidedly 'under the weather'.

by the breeze, it is worth bearing in mind, for the purposes of identification, that fog and mist are not strictly clouds at all, since they make direct contact with the ground – even though they are often described as varieties of ground-level Stratus, and they form in a similar way to stratiform clouds, when moisture-laden air touches cold ground or water at night, bringing the dew point (the temperature level at which water vapour condenses into liquid droplets) literally down to earth.

Stratus nebulosus comes in a variety of thicknesses, sometimes completely obscuring all light from above – at which the designation would be Stratus nebulosus opacus – and sometimes allowing the Sun or Moon to be seen with a clear outline, as can be seen in the atmospheric photograph on the previous page, of moonlight shining through a misty veil of Stratus nebulosus translucidus, to give it its full Latin name. Unlike its distant cousins, watery Altostratus and icy Cirrostratus (see C_M1 and C_H5 to C_H6), low Stratus clouds such as these do not produce optical phenomena such as haloes, parhelions and coronas, and neither are they capable of producing much in the way of rain.

Stratus clouds are often broken up by evaporation from a rising Sun, or the arrival on the scene of a layer of warm, turbulent air. When in the process of breaking upwards, Stratus clouds can sometimes appear more like low-lying cumuliform clouds, as seen in the photograph opposite, of a thick layer of Stratus nebulosus rising on convective currents, on their way to forming Stratocumulus stratiformis clouds.

Watching the behaviour of Stratus clouds over time can offer useful indications of short-term weather to come. If a layer of Stratus forms in air that is lifting over hill slopes, it may well be followed by rain, something that people who live in valleys know well. When low Stratus forms at night during the summer, the following morning can often start off gloomy, but the rising Sun will evaporate the water droplets, seeing off the early cloud, and leaving the rest of the day fine and clear.

OPPOSITE:
Stratus nebulosus
breaking up at 275m (900ft)
under the influence of rising
air currents.

C$_L$7

Stratus fractus

LEFT:
A windblown pocket of **Stratus fractus** (St fra) cloud collides into the hillside above Montreux, Switzerland.

OPPOSITE:
Wisps of **Stratus fractus** cloud swirl over Glencoe, Argyll, in advance of oncoming rain.

Stratus fractus are shreds of low cloud that can either appear separately, as in the photograph of a bank of white Stratus flailing over the lakeside town of Montreux in Switzerland, or beneath the base of another, precipitating cloud layer, such as Altostratus or Nimbostratus (C$_M$2), when they are collectively referred to as 'pannus' (see page 111).

Known also as 'scud' clouds, or 'messenger clouds', these ever-changing cloud fragments may well go on to merge and become a more or less continuous layer, sometimes even obscuring an extent of sky above. But usually they remain in the form of wisps of fast-moving cloud that breeze in just ahead of the rain, as in the photograph opposite, taken at Glencoe, in the Highlands of Scotland. The incoming cloud is not fog or mist, as it has clearly formed well above ground, but has swept horizontally into the side of the intervening hill, over which it will quickly pass to make way for the imminent rain that it heralds.

When Stratus fractus is accompanying a rainy cloud cousin such as Nimbostratus (C$_M$2), it can sometimes seem as if rain is falling from the lower Stratus cloud, when in fact the rain is falling through it from above. Stratus fractus itself rarely produces rain, and at most is capable only of short-lived bouts of drizzle.

C$_L$8

Cumulus and Stratocumulus at different heights

As described earlier, Stratocumulus clouds can form in two distinct ways: either by the spreading out of the upper parts of Cumulus congestus clouds, or from the breaking up of rising layers of Stratus nebulosus by upward convection. The clouds in this classification of sky are both of the latter (convective) variety, being small Cumulus clouds forming beneath an already present layer of Stratocumulus stratiformis, a relatively common situation that can clearly be seen in the two accompanying photographs.

As the rising Cumulus clouds approach the level of the Stratocumulus layer, they do not spread out horizontally, as is the case with Stratocumulus cumulogenitus clouds (C$_L$4) but tend to thrust their way through the upper layer, as is beginning to

RIGHT:
Rising **Cumulus** clouds thrust their way towards a layer of **Stratocumulus** 300m (1,000ft) above them.

happen in the first image, of Stratocumulus at c. 900m (3,000ft) being approached by Cumulus clouds c. 300m (1,000ft) below. Often the Stratocumulus layer will thin out or break up around the point of contact with the ascending Cumulus cloud.

Luke Howard described the changing face of this mixed-cloud sky in some detail in his 1803 essay *On the Modifications of Clouds*, observing how the upper cloud layer 'becomes denser and spreads, while the superior part of the Cumulus extends itself and passes into it.' The result of such mixing, as the photograph below (taken in Oklahoma) suggests, can be a complex and fascinating aerial landscape, which, 'seen as it passes off in the distant horizon, presents to the fancy mountains covered with snow, intersected with darker ridges and lakes of water, rocks and towers, &c.', as an evidently enchanted Howard concluded.

C$_L$9

SPECIFICATION: Cumulonimbus, the upper part of which is clearly fibrous (or cirriform), often in the shape of an anvil. This cloud may sometimes be accompanied by Cumulonimbus calvus (without anvil) (see C$_L$3).

Symbol =

Cumulonimbus capillatus

With its towering thunderhead, Cumulonimbus capillatus clouds can grow to vast heights, extending from their low bases, 600m (2,000ft) or lower, to 18km (60,000ft) or more above the ground – to where the troposphere meets the stratosphere – making them by far the tallest structures on Earth. No wonder the expression 'to be on cloud 9' (the classification number of Cumulonimbus that appeared in the first *International Cloud Atlas* in 1896) means 'to be on top of the world'.

The true scale of this enormous cloud is best appreciated when viewed against a flat horizon, as is the case in the dramatic photograph opposite of a vast single-cell Cumulonimbus structure, complete with its icy striated *incus* (Latin for 'anvil'), rearing over the small farming settlement of Dimmitt, Texas,

TOP:
Detail of the icy, striated anvil of a large **Cumulonimbus capillatus** (Cb cap) cloud.

RIGHT:
Cumulonimbus capillatus anvil rearing over Cape Town, South Africa.

OPPOSITE:
With as much energy as ten Hiroshima-sized atom bombs, storm-clouds such as this (over Dimmitt, Texas) can grow to vast heights, towering over the landscape below.

during a stormy summer afternoon. Thunder, lightning, hail, heavy rain and strong winds are the likely accompaniments to this vigorous cloud, which during its phase of maximum power can contain as much energy as ten Hiroshima-sized atom bombs.

Cumulonimbus clouds are usually an energetic development of Cumulus congestus clouds (see $C_L 2$), which have grown unstable through powerful upward convection currents. As was seen in the entry for Cumulonimbus calvus ($C_L 3$), which represents the intermediate phase between Cumulus congestus and the full-blown Cumulonimbus capillatus seen here, the characteristic cauliflower-shaped summit of the congestus cloud has been replaced by the glaciated anvil, a vast canopy of ice crystals which can sometimes be spread for hundreds of kilometres by the strong winds of the upper atmosphere, creating an eerie mushroom-cloud effect, as can be seen in the view on page 48 of a $C_L 9$ taken from Cape Town, South Africa.

From afar, the two Cumulonimbus species are unmistakeable, but viewed from immediately below the cloud base, where the entire sky is likely to be dark and lowering, and busy with pannus clouds (see page 111), both can easily be confused with Nimbostratus ($C_M 2$), the thick, grey blanket associated with persistent rain or snow, and unless it is actually hailing or thundery (the giveaway signs of a Cumulonimbus cloud), it is the quality of the rain that will assist identification: widespread, persistent rain is most likely to fall from a Nimbostratus cloud, while sudden, heavy bursts or sharp showers are more likely to come from a Cumulonimbus. From immediately below a Cumulonimbus cloud, of course, the top cannot be seen, making identification of the species (calvus or capillatus) impossible, unless you telephone someone in a nearby town who can see the cloud from the side … In that situation, the classification $C_L = 9$ is used.

Individual Cumulonimbus clouds, of the kind shown here, are relatively short-lived single-cell varieties, lasting perhaps an hour or more before raining or blowing themselves out. If a number of single-cell Cumulonimbus clouds manage to coalesce, however, they can form a multicell or even a supercell storm structure that can last for many hours, or even for days, and it is these great storm systems, with their extreme vorticity, that are responsible for producing the thousand or more tornadoes that batter the American Midwest every year.

OPPOSITE:
The end of the stormcloud: a large
Cumulonimbus capillatus cloud
passes over farmland in Kansas, its
volume rapidly diminishing through
heavy precipitation.

Medium Clouds

(of genera Altostratus, Nimbostratus and Altocumulus)

C$_M$ 1

> **SPECIFICATION:** Altostratus, the greater part of which is semi-transparent; through this part the Sun or Moon may be weakly visible, as through frosted glass.
>
> Symbol = ∠

Altostratus translucidus

Altostratus translucidus is a featureless layer of thin, grey-blue cloud that can spread to cover most of the sky, giving rise to dull, overcast conditions. It can form from a descending veil of Cirrostratus (see C$_H$5), as well as, less commonly, from the spreading out of the upper parts of a Cumulonimbus cloud, but it is usually the result of the lifting of a large mass of warm air ahead of an incoming warm front or occlusion. If the warm front continues to advance, pushing further moist air upwards, Altostratus translucidus can thicken into Altostratus opacus or Nimbostratus (see following entry), a sure sign of imminent rain. When the sky is filled with thin Altostratus, sunlight becomes diffused and watery, and rarely casts a shadow on the ground. A rising or setting Sun will glow pink or orange through the cloud veil, as seen in the image of a Bracknell sunset above, while optical phenomena such as coronas and iridescence are common accompaniments (see pages 144 and 128), due to the cloud being largely made up of uniformly sized water droplets.

ABOVE:
A sunset in Bracknell, Berkshire, seen through a thin layer of **Altostratus translucidus** (As tr).

OPPOSITE:
Formed by the lifting of a large mass of warm air, **Altostratus translucidus** clouds form a dull, featureless layer across the sky.

C_M2

> **SPECIFICATION:** Nimbostratus is a thickened, rainy form of Altostratus opacus (also C_M2), the greater part of which is dense enough to hide the Sun or Moon from view.
>
> Symbol =

Altostratus opacus, or Nimbostratus

With the exception of gardeners at the height of a hosepipe ban, it is a safe bet that Nimbostratus would turn out to be everybody's least favourite kind of cloud. Grey, gloomy and pouring with rain, the cloud blanket that seems to last all day (especially at weekends and at major sporting fixtures) usually begins life as a high, thin layer of Altostratus translucidus (see previous entry) which descends and thickens as it takes on rising moisture imported by an advancing warm front. As the cloud thickens, sometimes in several layers, the Sun or Moon will become increasingly obscured, and the Altostratus translucidus variety will be redefined as an Altostratus opacus, as shown in the photograph on the right, in which weak sunlight struggles to appear behind the thickening veil of cloud. Even if the cloud layer grows so thick that the Sun or Moon is

RIGHT AND OPPOSITE: Grey, dull and depressing, **Nimbostratus** (Ns) is probably the world's least favourite cloud. It forms from thickened **Altostratus opcaus** (As op) clouds, as seen in the image on the right.

screened out entirely, the cloud is not yet technically a Nimbostratus until rain starts to fall from it (*nimbus* being the Latin word for 'rain cloud').

Nimbostratus is unique among the ten cloud genera for having no species or varieties to its name, although it can occur at a range of altitudes, despite the fact that it is classified as a medium-level cloud. The second image, taken at Bracknell in Berkshire, is of a relatively high Nimbostratus, with its raggedy,

grey base some 2.75km (9,000ft) above the ground, although usually the cloud base would be well below 2km (6,500ft), sometimes lowering only a few hundred feet above the ground. At such times, with near-continuous sheets of rain, snow or sleet serving to diffuse any remaining light, it can appear to be connected directly with the sodden landscape below, leaving us feeling not just under the weather, but in it.

Both Altocumulus opacus and Nimbostratus can

be very long-lasting clouds, and over time the air below them can become highly saturated, which is when ragged shreds of pannus cloud develop beneath the main cloud layers, as can be seen in the image opposite, a gloomy shot taken through falling rain on a near-lightless day. During heavier bouts of rain or snow, the pannus clouds are often dispersed, but they tend to stage a return during lulls in the precipitation.

RIGHT:
A ragged band of **Altostratus** opacus, thickening into rainy **Nimbostratus**.

OPPOSITE:
A gloomy day in Berkshire: almost no light penetrates the thick layer of **Nimbostratus** that extends as far as the eye can see. Some shreds of **Stratus fractus** can also be seen scurrying below.

C_M3

SPECIFICATION: Altocumulus, the greater part of which is semi-transparent; the various elements of the cloud change only slowly and are all at a single level.

Symbol = ∽

Altocumulus translucidus and Altocumulus stratiformis at a single level

Altocumulus clouds are a group of medium-level cloud species that can occur as rounded, individual 'bread-roll' clouds, either in localized patches or in vast, extensive layers, as well as in haunting 'UFO' formations (see Altocumulus lenticularis, C_M4). They can form in a variety of ways, such as the slow breaking up (due to gentle convective currents) of a stratiform cloud layer such as Altostratus, or from pockets of moist air that are lifted and cooled by gentle turbulence to form a mid-level layer of cloud. In contrast to the strong local convection currents that are responsible for forming lower-level Cumulus clouds, the turbulent currents that create Altocumulus tend to undulate gently in the upper air, like waves in water. As each wave rises, water vapour condenses, forming isolated patches of visible cloud, but where

RIGHT:
A radial example of **Altocumulus stratiformis** (Ac str) in the sky above Norman, Oklahoma, formed from the breaking up of Altostratus clouds.

OPPOSITE:
Altocumulus stratiformis clouds arranged into parallel bands.

the wave sinks or troughs, gaps are created between the individual elements. Altocumulus clouds are usually composed of either supercooled water droplets or ice crystals, or a mixture of both, so they are equipped to exhibit a wide range of optical phenomena, depending on which form of water is predominant (see the entries on optical phenomena on pages 140–153).

Altocumulus clouds are so varied in their arrangements that they have come to occupy seven of the 27 categories of sky, this first of which features semi-transparent cloudlets in a single layer, not progressively invading the sky. Despite their variant formations – radial, fleecy and cellular – all clouds of C_M3 exhibit the same degree of translucency, and all can clearly be seen to persist as a single layer of cloud, in which the individual elements change very little over time.

RIGHT AND OPPOSITE:
Altocumulus stratiformis can form in rounded 'bread-roll' cloudlets, such as those shown on this page, or as a single layer of fleecy patches, such as the sky above Wester Ross, in the Highlands of Scotland (opposite).

C_M4

SPECIFICATION: Patches of Altocumulus, often lenticular ('lens- or almond-shaped'), the greater parts of which are semi-transparent; these clouds occur at more than one level, and are continually changing in appearance.

Symbol = 𝒞

Altocumulus lenticularis

Lenticular ('lens-shaped') Altocumulus clouds are formed when a flowing layer of moist air is uplifted by the slope of an intervening hill or mountain. The wind that carries the air over a mountain rises gently, cooling unevenly as it does so, sending bouncing waves of moisture-laden air streaming away from the obstruction. Lenticular wave clouds form in the crests of these air waves, and dissipate in the troughs, appearing to hover in a fixed position while a current of air streams through them, condensing its moisture at one end of the cloud, and evaporating it at the other. Such standing wave clouds can occur at considerable distances from the mountains themselves, forming repeating patterns of regular lenticular cloud lines.

RIGHT:
A heart-shaped
Altocumulus lenticularis
(Ac len) formed by a
layer of moist air uplifted
over the Shropshire hills.

OPPOSITE:
UFO-shaped lenticular
clouds over Aberdeen.
Known to gliders as
'lennies'.

These beautiful undulating clouds often emerge and dissipate in unexpected ways, according to the movement of the air currents, sometimes appearing UFO-shaped, as in the striking example on the previous page photographed over Aberdeen, or just vaguely monstrous, as in the cloud photographed rearing over Raisbeck, in Cumbria. Sometimes they appear in stacked layers, like the formation below, seen above the Isle of Islay, in Scotland, a circumstance known as a *pile d'assiettes*, from the French for 'stack of plates' (a rare example of cloud terminology deriving from a language other than Latin or Greek). They are always localized phenomena, however, and never go on to invade the sky like their near relatives Altocumulus stratiformis (see following entry, C$_M$5). These elegant, shape-shifting lenticulars are surely the clouds that Shakespeare had in mind when he had prince Hamlet tease the courtier Polonius in their famous exchange:

HAMLET: Do you see yonder cloud that's almost in shape of a camel?
POLONIUS: By th' mass, and 'tis like a camel, indeed.
HAMLET: Methinks it is like a weasel.
POLONIUS: It is backed like a weasel.
HAMLET: Or like a whale.
POLONIUS: Very like a whale.

(*Hamlet*, Act III, scene ii)

RIGHT:
UFO-shaped lenticular clouds in an eerie *pile d'assiettes* ('stack of plates') formation, Isle of Islay, Scotland.

OPPOSITE:
A Zeppelin-shaped **Altocumulus lenticularis** cloud over Raisbeck, Cumbria.

C_M5

> **SPECIFICATION:** Semi-transparent Altocumulus, either in bands, or in one or more fairly continuous layers, which progressively invade the sky, growing thicker as they do so.
>
> Symbol =

Altocumulus stratiformis

Although similar in appearance to the clouds of C_M3 (Altocumulus stratiformis at a single level) this is another example of how cloud classification takes transformation over time to be one of its identifying characteristics. Still photographs cannot really do justice to the distinction between clouds of the C_M3 coding, and those of the C_M5, but the former are much more static and localized, while the latter progressively invade the sky, gathering overhead rapidly, and thickening as they do so, sometimes creating an entirely covered sky that stretches from horizon to horizon, as can be seen in the beautiful evening photograph opposite. These clouds often progess in parallel bands that spread through gentle turbulence into a variety of attractive formations. Their advancing edges may also consist of small

RIGHT:
Altocumulus stratiformis (Ac str) moving across the sky, ripples forming in the gentle turbulence.

OPPOSITE:
An evening sky over Tyneside, filled with **Altocumulus stratiformis** cloudlets, which noticeably thicken as they spread across the horizon.

cloudlets that progressively thicken, as can be seen in the photograph of a band of Altocumulus moving over Ebbw Vale in Gwent (opposite).

This thickening is often due not to the cloudlets themselves growing in size, but to the increased concentration of water droplets present within them, lending them an 'optical depth' that is sometimes enough to completely obscure the Sun or Moon. These clouds are worth keeping an eye on, since they, or at least parts of them, are liable to thicken and descend, transforming into Altostratus opacus clouds, or even Nimbostratus, at which the coding would change to C_M7 and then, depending on what happens next, to a rain-sodden Nimbostratus (C_M2).

RIGHT:
A dappled sky filled with thickening cloudlets of **Altocumulus stratiformis**.

OPPOSITE:
The advancing edge of a band of **Altocumulus stratiformis** moving over Ebbw Vale, in Gwent. The cloudlets progressively thicken as they move across the sky.

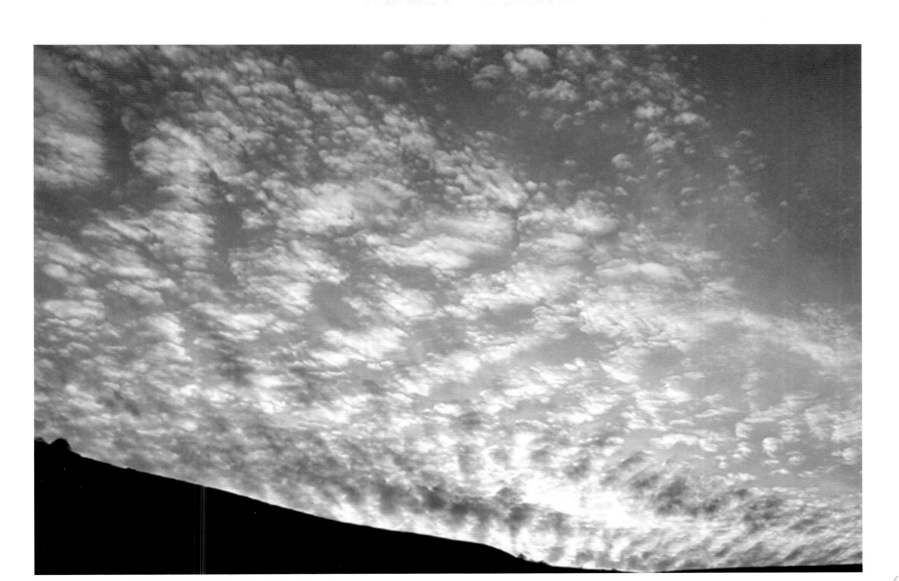

C_M6

SPECIFICATION: Altocumulus resulting from the spreading out of Cumulus (or Cumulonimbus) clouds.

Symbol =

Altocumulus cumulogenitus

LEFT:
Altocumulus cumulogenitus
(Ac cugen) in its early stage of formation, in which patchy cloudlets can be seen spreading from the top of a **Cumulus congestus** cloud.

OPPOSITE:
Altocumulus cumulogenitus
spreading laterally, in the shape of an anvil. The cloud's upward growth has been arrested by a temperature inversion in the upper atmosphere.

As is also the case with Stratocumulus cumulogenitus clouds (C_L4), the upward growth of cumuliform clouds can be halted by the presence of a temperature inversion, at which they begin to spread out horizontally instead, to form an entirely new species of cloud. An early stage of this process can be seen in the photograph on the left, in which patchy cloudlets of Altocumulus can be seen spreading out from the top of a Cumulus congestus cloud. If the spreading continues, the patches will tend to thicken and join, creating an extensive wedge-shaped layer of cloud. Although Altocumulus cumulogenitus can sometimes appear in the shape of an anvil, it never has the striations or the icy sheen of a true Cumulonimbus anvil, as seen in the summits of C_L9 itself (see page 48).

Sometimes, the spreading of this cloud is only temporary, and upward growth is soon resumed, so that the Altocumulus appears to one side of the new, higher Cumulus cloud.

This species of Altocumulus can also be formed from the elements left behind by a decayed Cumulonimbus cloud, a formation for which the designation would more correctly be Altocumulus cumulonimbogenitus (Ac cbgen), although the coding would still be C_M6.

C_M7

> **SPECIFICATION:** Altocumulus translucidus, stratiformis or opacus in two or more layers, not progressively invading the sky; or a single layer of Altocumulus opacus or Altocumulus stratiformis, not progressively invading the sky; or Altocumulus appearing with Altostratus and/or Nimbostratus.
>
> Symbol =

Altocumulus stratiformis duplicatus, or Altocumulus with Altostratus or Nimbostratus

The specification C_M7 is used to describe three closely related varieties of sky:

a) patches or sheets of Altocumulus stratiformis in more than one layer (duplicatus), as shown in the photograph opposite of a dramatic evening sky over Bracknell, in Berkshire, in which two distinct layers of Altocumulus stratiformis (recognizable from the C_M5 category) have formed, one at 2.4km (8,000ft), the other at 3.6km (12,000ft). Such a combination of cloud layers, even if they are individually quite thin and diffuse, can sometimes prove dense enough to mask the Sun or Moon completely, and as these clouds do not change continually, nor do they invade the sky, they can be quite persistent, lasting until sunset.

RIGHT:
A localized patch of thick **Altocumulus stratiformis** (Ac str) formed over Swinley Forest, Berkshire.

OPPOSITE:
Altocumulus stratiformis duplicatus (Ac str du) in more than one layer, creating a richly patterned evening sky, with the upper layer lit from below, contrasting with the deeply shadowed bottom layer, 1.2km (4,000ft) below.

A sky state much like this was described by the cloud-haunted Jesuit poet Gerard Manley Hopkins in a journal entry for 23 July 1874:

'It was a lovely day: shires-long of pearled cloud under cloud, with a grey stroke underneath marking each row.'

b) thick Altocumulus occurring in a single layer, a particularly beautiful example of which can be seen in the photograph on page 72. Clouds like these often occur in localized patches, and do not progressively invade the sky; their cloudlets can bear a superficial resemblance to those of Cirrocumulus floccus (C_H9), although Altocumulus elements will tend to exhibit three-dimensional shading; any doubt can usually be dispelled by holding up one's hand at arm's length, and measuring the width of a cloudlet with one's fingers: those of Altocumulus will typically measure two or three fingers' wide, while those of Cirrocumulus will usually only measure one.

c) an occurrence of Altocumulus in tandem with Altostratus, or even Nimbostratus, as seen in the image below of a cloudy sky over the Isle of Skye, as well as in the photograph opposite. This sky results from local transformation processes, through which Altocumulus clouds acquire the appearance of Altostratus. These clouds can appear in two or more layers, with each layer showing certain characteristics of both species, as in the Isle of Skye image, in which a more stratiform layer rides over a slightly more cumuliform layer. Another example of this can be seen in the centre of the photograph opposite.

RIGHT:
Altocumulus with **Altostratus**, in distinct layers, their bases between 3 and 4.5km (10,000 and 15,000ft), Isle of Skye from Mallaig, Scottish Highlands.

OPPOSITE:
Altocumulus with **Altostratus** Bracknell, Berkshire.

C_M8

SPECIFICATION: Altocumulus with sproutings
either in the form of towers or castellations
(castellanus) or small cumuliform tufts (floccus).
Also roll cloud (volutus).

Symbol = M

Altocumulus castellanus; Altocumulus floccus; Altocumulus volutus

The cloudlets that make up Altocumulus castellanus formations exhibit sproutings in the form of small towers or battlements, a sure sign of instability in an upper layer of the sky. The cloud elements themselves have a common base and sometimes appear to be arranged in lines, as can be seen in the photograph, on the right, of a wind-ordered file of these clouds marching against an early evening sky. The larger the castellations, the more vigorous the instability, and a sighting of these clouds can be a reliable indication of thunderstorms coming in over the next 24 hours. Altocumulus castellanus can also descend and merge to form large Cumulus (C_L2) or sometimes even Cumulonimbus clouds (C_L3 or C_L9), a process that could be in its early stages in the photograph opposite, taken in northern France.

Altocumulus floccus (which can sometimes form

RIGHT:
An orderly file of
small **Altocumulus
castellanus** (Ac cas)
clouds marching across
an early evening sky
in Berkshire.

OPPOSITE:
Altocumulus castellanus
descending and
merging to create
larger cumuliform
clouds, Mascent,
northern France.

from the dissipated bases of Altocumulus castellanus) appear as white or grey scattered tufts with rounded and slightly bulging upper parts, resembling small ragged Cumulus clouds. They often feature fibrous trails of virga (rain or snow not reaching the ground, see page 116) trailing from their bases.

Like its close relative Altocumulus castellanus, the floccus species of Altocumulus is also associated with humid, unstable conditions, likely to lead to thundery conditions developing over a wide area (as opposed to local thunderstorms originating from Cumulonimbus clouds immediately overhead).

An evening display of them will often precede wet or stormy weather in the morning, especially if convection kicks in with the rising Sun, and young, energetic Cumulus clouds end up joining in with the moisture-laden Altocumulus clouds already present.

The rarest of all the Altocumulus clouds is volutus (or 'roll-cloud'), which takes the form of a detached, tube-shaped cloud mass that often appears to roll slowly about a horizontal axis. Caused by differences in wind speed and direction between the base of the cloud and its summit, Ac vol usually occurs as a single line and seldom extends from horizon to horizon.

RIGHT:
Tufts of **Altocumulus floccus** with **virga** (Ac flo vir) – fallstreaks of icy rain or snow which evaporate before reaching the ground.

OPPOSITE:
Such evening displays of **Altocumulus floccus** will often precede wet or stormy weather in the morning.

C_M9

SPECIFICATION: Altocumulus of a chaotic sky, generally at several levels.

Symbol =

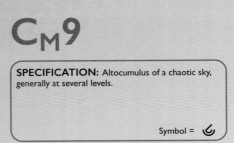

Altocumulus of a chaotic sky

Altocumulus of a chaotic sky usually occurs at several levels. The sky is characterized by a heavy appearance, with broken sheets of poorly defined clouds of several transitional stages of medium-level clouds, from low, thick Altocumulus, to high, thin Altostratus. The weather prognosis is uncertain when the sky looks like this, although anyone seeing it would be well advised to keep an umbrella handy.

TOP, RIGHT AND OPPOSITE: In the first of this sequence of images of **Altocumulus** (Ac) of a chaotic sky over Bracknell in Berkshire, a variety of cloud forms swirl around together, a turret of Altocumulus castellanus even making an appearance towards the right of the picture. In the image opposite, clouds of many varieties appear at several levels.

High Clouds
(of genera Cirrus, Cirrostratus and Cirrocumulus)

C_H1

C_H1

> **SPECIFICATION:** Cirrus clouds in the form of filaments, strands or hooks, not progressively invading the sky.
>
> Symbol =

Cirrus uncinus; Cirrus fibratus

The high, white, delicate Cirrus cloud species of C_H1 occur generally in curved filaments or straight lines, 'pencilled, as it were, on the sky', as Luke Howard described their appearance in 1803. Like all Cirrus clouds, they are composed entirely of gently falling ice crystals, at altitudes above c. 6km (20,000ft). They sometimes form from the virga (see page 116) of high Cirrocumulus clouds, but are usually formed when layers of relatively dry air ascend in the upper troposphere, the small amount of vapour then subliming into ice when it meets its subzero dew point (sublimation is the process of transforming directly from a solid to a gas, or vice versa, with no intermediate liquid state). These smaller kinds of lofty Cirrus cloud usually appear on their own in a dry

RIGHT AND OPPOSITE:
Composed entirely of
gently falling ice crystals,
the curved filaments of
Cirrus uncinus (Ci unc)
form easily identified
commas in the sky.

blue sky, since if the air were more humid, other types of cloud would have formed at lower levels. If that is the case, and the Cirrus does not begin to spread – as is the situation in the first two photographs – then fine weather may well continue for a while, but if the Cirrus begins to increase its cover, thickening or spreading out horizontally, then it means that a warm front is on its way, pushing up moist air ahead of itself, and causing the weather to take an imminent turn for the worse. Sailors have long viewed the growth of comma-shaped Cirrus clouds as a useful 'wind warning', and in the photograph below right, the crystals falling from a group of 'mares' tails' (a popular name for Cirrus uncinus) can be seen being drawn into long filaments by the wind: 'mares' tails and mackerel scales make tall ships carry low sails', as the old weather-motto had it.

Cirrus fibratus can often be arranged by the wind into parallel bands, as can be seen in the photograph opposite, which appear to converge towards the horizon. Even in such relatively dense arrangements, cirriform clouds still appear wispy and diffuse compared to the majority of low and medium cloud formations, because the concentrations of ice crystals in Cirrus clouds are so much lower than the droplet concentrations found in clouds that are composed of liquid water.

RIGHT:
The mariner's wind-warning: 'Mares' tails' of **Cirrus uncinus** being drawn into long filaments by the wind. The longer the filaments grow, the stronger the wind will become.

OPPOSITE:
Cirrus fibratus (Ci fib) clouds arranged into bands by the winds of the upper atmosphere, Charmouth, Dorset.

C_H2

> **SPECIFICATION:** Dense Cirrus, in patches or sheaves, which usually do not increase (spissatus); or Cirrus with sproutings either in the form of small turrets or battlements (castellanus) or small cumuliform tufts (floccus).
>
> Symbol =

Cirrus spissatus; Cirrus castellanus; Cirrus floccus

Cirrus spissatus is a thick, dense species of Cirrus cloud, which can often appear to dominate large areas of sky. Such thickened or actively changing Cirrus species will often appear in the vanguard of an approaching warm front, as a parcel of moist air is forced up over the shallow wedge of cooler air that it encounters, causing high clouds to form in the upper atmosphere. It often means that wet or stormy weather will be on its way some time over the following 48 hours.

The castellanus and floccus species that are also classed as C_H2 exhibit turretty sproutings or ragged patches, and often feature trailing filaments below the main cloud, as can be seen to spectacular effect in the photograph of Cirrus castellanus cloudlets, with

RIGHT AND OPPOSITE: In these two photographs taken in Devoran, Cornwall, icy **Cirrus spissatus** (Ci spi) clouds grow to dominate a large area of sky.

what look like lengthy striated kite-tails hanging below. These tendrils form when the descending ice crystals end up in a deep layer of cold air that is moving at a stable speed, causing them to be spread across the sky, sometimes for enormous distances. Although these Cirrus clouds spend their careers slowly falling, and are thus technically 'precipitating clouds', they rarely produce any kind of precipitation that reaches the ground, although they often exhibit virga (see page 116), trails of descending snow or rain that evaporate in the warm lower air, long before they are able to reach the surface.

Cirrus clouds can also form from the spreading out of aircraft contrails, which grow by seeding moisture present in the air, and which can sometimes last for many hours as the special cloud Cirrus homomutatus, becoming almost indistinguishable from naturally occurring clouds (see page 138).

RIGHT:
High **Cirrus castellanus**
(Ci cas) cloudlets trailing
virga (fallstreaks) from
their bases.

OPPOSITE:
Fleecy cloudlets of
Cirrus floccus (Ci flo),
Bracknell, Berkshire.

C$_H$3

SPECIFICATION: Dense Cirrus, often in the form of an anvil, being the remains of the upper parts of a Cumulonimbus cloud.

Symbol = ⌐

Cirrus spissatus cumulonimbogenitus

LEFT:
The leftover anvil of a rained-out Cumulonimbus cloud, **Cirrus spissatus cumulonimbogenitus** (Ci spi cbgen) is nevertheless a cloud species in its own right.

OPPOSITE:
Cirrus spissatus cumulonimbogenitus, seen from Grand Anse Beach, Grenada.

This form of Cirrus derives from the leftover anvil of a decayed or rained-out Cumulonimbus capillatus cloud (C$_L$9) (see also Incus, page 114). The Cumulonimbus anvil is essentially a vast canopy of turbulent Cirrus that can sometimes be carried away from the main stormcloud by the high-speed winds of the upper atmsophere, as has clearly happened in the first photograph (left) of a marooned anvil flailing above Reading, in Berkshire. The resultant C$_H$3 Cirrus clouds are usually frayed and battered at the edges, but are still sufficiently thick to veil the Sun, and therefore grey in colour, in contrast to the characteristic delicate whiteness of Cirrus uncinus (C$_H$1) or Cirrus castellanus (C$_H$2) clouds.

Other cirriform clouds may also be present at the same time as C$_H$3, so it is sometimes hard to be absolutely certain that the cloud is truly cumulogenitus, rather than merely an energetic example of Cirrus spissatus (see previous entry). In uncertain situations such as this (as shown in the photograph opposite) – especially if one has not actually witnessed the transformation of the anvil at first hand – it is perfectly acceptable to use the simpler Cirrus spissatus (C$_H$2) designation.

C_H4

> **SPECIFICATION:** Cirrus in the form of hooks (uncinus) or filaments (fibratus) (see clouds of C_H1) progressively invading the sky, generally thickening as they do so.
>
> Symbol = $\supset$

Cirrus uncinus or Cirrus fibratus progressively invading the sky

The Cirrus clouds of C_H4 are the same species as those of C_H1, but with the added feature of their progressively invading the sky, due to the influence of an approaching warm front as it serves to slide expanses of warm moist air over wedge-shaped areas of colder air, causing icy clouds to form at high altitudes (around 6km/20,000ft). When Cirrus clouds are seen to thicken and increase, sometimes spreading to entirely cover the sky, it is a sure indication of an advancing depression, an area of low pressure and rising air that is usually associated with imminent bad weather.

These clouds often seem to fuse together in the direction from which they first appear, as in the main photograph of spreading Cirrus fibratus taken in Pontypool, in Gwent, but they can also arrange themselves into great rippling parallel formations, as

RIGHT:
'Mares' tails and mackerel scales make tall ships carry low sails': these characteristic mares' tails of **Cirrus uncinus** (Ci unc) invading the sky over Padstow, Cornwall, are a sure sign of turbulent weather to come.

OPPOSITE:
Bad weather on its way: **Cirrus fibratus** (Ci fib) invading the sky, in advance of an approaching depression, Pontypool, Gwent.

in the stunning example of invading Cirrus opposite, taken at Freshwater, Isle of Wight. Ice crystal-related optical phenomena such as haloes, mock Suns and circumzenithal arcs (see pages 140–143) are occasional accompaniments to this attractive species of sky, and serve as further indications of the deteriorating weather conditions to come. They will often thin out over time to form uniform veils of Cirrostratus (see following entries).

RIGHT:
Dense formations of **Cirrus uncinus** fill the evening sky, promising a bout of bad weather in the morning.

OPPOSITE:
Cirrus fibratus clouds can sometimes arrange themselves into parallel bands, as in this example seen over Freshwater, Isle of Wight.

C$_H$5

Cirrostratus (below 45°)

LEFT AND OPPOSITE: The leading edges of these bands of **Cirrostratus** (Cs) can clearly be seen as they invade the sky below 45°.

Cirrostratus clouds are high ice-crystal clouds produced by the slow ascent of air in the troposphere. Like Cirrus clouds, from which they often evolve, they are formed when water vapour sublimes into ice crystals, often ahead of advancing weather fronts, at altitudes above 6km (>20,000ft). The movements of Cirrostratus clouds are always worth keeping an eye on, because their behaviour can often give advance notice of changes in the weather to come. If it is seen to form out of spreading Cirrus (C$_H$4) that grows thicker and more continuous, it is likely to be followed by wet weather within 48 hours, but if gaps begin to appear in the Cirrostratus veil, and it begins slowly to change into Cirrocumulus (C$_H$9), the weather will probably continue to be dry for a day or so (but remember to keep an eye on the developing Cirrocumulus).

Cirrostratus formations of C$_H$5 and C$_H$6 sometimes have a definite 'edge' to them, as can be seen in both these photographs, although they can also be fringed with Cirrus clouds, making initial identification more difficult. Skies of C$_H$5 will often feature bands or tufts of thickened Cirrus appearing ahead of the Cirrostratus, but the principal component is always an advancing veil of whitish Cirrostratus slowly advancing over the horizon, but (at least in this classification) no more than 45° above it.

C_H6

> **SPECIFICATION:** Cirrus, often in bands,
> with Cirrostratus, or Cirrostratus on its own,
> progressively invading the sky, and growing denser
> as it does so, with the continuous veil of cloud
> extending more than 45° above the horizon, although
> without the sky being totally covered.
> Symbol = *2*

Cirrostratus (above 45°)

If the Cirrostratus of C_H5 advances further, becoming
continuous to more than 45° above the horizon,
but without covering the entire sky, its appearance is
coded as C_H6 (although the distinction between C_H5
and C_H6 may seem a little academic for the purposes
of everyday cloud identification). But the cloud itself
changes as it advances, tending to grow denser as
it proceeds, revealing a more fibrous structure, and
with a less clear-cut leading edge, as can be seen in
these photographs here. As with the sky of C_H5, it is
often accompanied by neighbouring Cirrus clouds,
examples of which can easily be seen in the main
photograph opposite.

RIGHT:
Cirrostratus,
(Cs) invading the
sky and above 45°
elevation, its leading
edge now less clear-cut
than in the preceding
classification.

OPPOSITE:
Cirrus clouds
accompany this
spreading band
of **Cirrostratus**, as it
progressively invades
the sky.

C_H7

SPECIFICATION: A veil of Cirrostratus which covers the entire sky.

Symbol =

Cirrostratus nebulosus or Cirrostratus fibratus invading the whole sky

Following on from the previous installments of Cirrostratus advancing across the sky in stages, the sky of C_H7 is one entirely covered by a high-riding veil of Cirrostratus, which can sometimes be quite dense and fibrous in appearance (Cirrostratus fibratus, which is shown in the photograph above right), and at other times so thin that it goes completely unnoticed, having little or no direct effect on sunlight (this is Cirrostratus nebulosus, and is shown opposite, in the image of sunlight being gently diffused through a dappled gauze of cloud). Shadows, for example, can still be cast from light shining through Cirrostratus, in contrast to its lower-level counterpart, the denser Altostratus cloud (see C_M1, page 52). At times the thinner species of Cirrostratus can appear so milky and indistinct that the only clue that they are there at all is the sight of a halo around the Sun or Moon, an effect caused by light refracting through the clouds' hexagonal ice crystals (see page 140 for a more detailed description of halo phenomena).

This kind of Cirrostratus, however, is prone to thicken and descend at the onset of a warm front, transforming by degrees into Altostratus (C_M1), which is often a precursor of Nimbostratus (C_M2), the soggy culmination of the complex, day-long layer-cloud cycle that so often begins with the first appearance of the high, delicate mares' tails of Cirrus uncinus (C_H1).

ABOVE:
A dense layer of **Cirrostratus fibratus** (Cs fib) that has advanced to cover much of the evening sky.

OPPOSITE:
A thin veil of **Cirrostratus nebulosus** (Cs neb) through which dappled sunlight can clearly be seen.

C$_H$8

SPECIFICATION: A veil of Cirrostratus which neither covers the entire sky, nor progressively invades it.

Symbol = ⌐c

Cirrostratus not progressively invading the sky

The sky state categorized as C$_H$8 features a veil of Cirrostratus that is not, or is no longer, invading the sky progressively, and which does not cover it entirely, as distinct from the invasive categories of Cirrostratus outlined earlier (C$_H$5 to C$_H$7). It typically appears as a broken patch of Cirrostratus fibratus, as in the two photographs on the right, with edges that are either clear-cut or fairly ragged in appearance, although Cirrus or Cirrocumulus might also be present, without predominating over the main body of Cirrostratus.

TOP AND RIGHT: These two photographs show patchy **Cirrostratus** clouds neither invading the sky nor entirely covering it.

OPPOSITE: **Cirrostratus,** in all its manifestations, has often lent its services to the creation of colourful sunsets, as can be seen in this dramatic example.

C$_H$9

SPECIFICATION: Cirrocumulus appearing alone, or with Cirrus and/or Cirrostratus, as long as the Cirrocumulus is predominant.

Symbol = $\mathcal{2}$

Cirrocumulus stratiformis; Cirrocumulus floccus; Cirrocumulus lenticularis

Cirrocumulus clouds are a group of relatively rare cloud species, made up of ice crystals and supercooled water droplets ('supercooled' refers to water that remains in a liquid state at temperatures well below freezing), which occur either in small, rippled, 'grainy' white patches or in shallow, extensive formations across the sky. They form high in the atmosphere, between c. 5 to 14km (16,000 and 45,000ft) when turbulent upward air currents encounter high Cirrus or Cirrostratus clouds, transforming some of their ice crystals into supercooled water droplets, while breaking them up into the rounder cloudlets of Cirrocumulus. Luke Howard recorded his detailed observations of this process in his 1803 essay: 'the Cirrocumulus is formed from a Cirrus, or from a number of small separate cirri, by the fibres collapsing, as it were, and passing

TOP AND RIGHT:
Small patches of 'herringbone' **Cirrocumulus stratiformis** (Cc str), made up of ice crystals and supercooled water droplets.

OPPOSITE:
Strong upward convection is responsible for the fleecy appearance of these **Cirrocumulus floccus** (Cc flo) cloudlets.

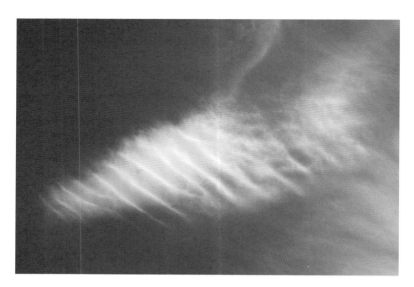

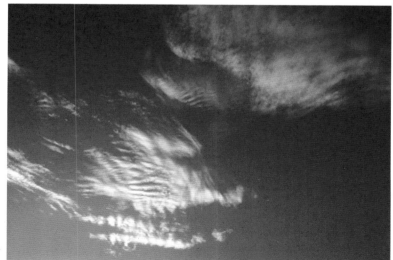

into small roundish masses, in which the texture of the Cirrus is no longer discernible, although they retain somewhat of the same relative arrangement.' This vestigial resemblance to Cirrus is evident in the top photograph on page 104, of a patch of 'herringbone' Cirrocumulus stratiformis seen high in the sky above the Lleyn Peninsula in North Wales.

Cirrocumulus floccus form when the convection that gives rise to the cloudlets is stronger and more unstable, serving to emphasize the cumuliform structure of the new cloud formation, despite its cirriform origins. In hilly areas, by contrast, the effects of orographic wave motion through a moist, stable layer of air high in the sky can produce the rare Cirrocumulus lenticularis formation, as seen in the photograph on the right – a cloud not to be confused with the lower-lying, and slightly more common Altocumulus lenticularis (C_M4).

Due to the generally unstable conditions in which they form, these smaller appearances of Cirrocumulus tend to be short-lived, either thinning and spreading out to form veils of Cirrostratus, or joining up with others to cover large areas of sky, as can be seen in the main photograph opposite. It is sometimes difficult to distinguish these larger outbreaks of Cirrocumulus from their lower Altocumulus cousins, although it should be noted that the cloudlets of Cirrocumulus tend to look much smaller than those of Altocumulus stratiformis (C_M5), since they are so much further away, while their contours are never shaded in the way that the thicker elements of

Altocumulus formations usually are. If in doubt, one easy test is to hold up a hand at arm's length and spread it against the sky – Cirrocumulus cloudlets are rarely more than one finger-width across, whereas Altocumulus cloudlets tend to be two or even three times that size.

Broad, rippled displays of Cirrocumulus stratiformis are often referred to as 'mackerel skies', due to the cloudlets' resemblance to fish scales. Mackerel skies have long been known as harbingers of bad weather, especially by seafarers, whose ancient saying: 'mares' tails and mackerel scales make tall ships carry low sails', is a testament to the fact that a large amount of moisture borne so high in a cold sky is a

visible indication of an advancing depression, while the broken-up quality of the cloud layer itself shows that conditions up there are turbulent.

In the course of researching his humorous *Cloudspotter's Guide* (2006), Gavin Pretor-Pinney (founder of the Cloud Appreciation Society) paid an early morning visit to Billingsgate Fish Market in East London, with the aim of identifying the species of mackerel that a mackerel sky most closely resembles. It is, he reported, the migratory King mackerel (*Scomberomorus cavalla*), the skin of which is patterned with precisely the same bands of silvery-white ripples that often appear high in the sky before the onset of stormy weather.

LEFT:
Cirrocumulus lenticularis (Cc len), Painswick, Gloucestershire.

OPPOSITE:
Cirrocumulus stratiformis (Cc str) spreading out to cover a large area of sky over Tynedale, Northumberland. Note the 'distrail' (the opposite of a contrail), cut through the cloud by a passing aeroplane.

PART 2:
OTHER CLOUDS
AND EFFECTS

Accessory Clouds

Pileus (pil)

From the Latin for 'cap', pileus is a supplementary layer of flattened cloud that sometimes appears above a Cumulus congestus (C_L2) or Cumulonimbus calvus (C_L3) cloud. It is formed by the rapid condensation of a layer of moist air that is pushed up over the cloud's main summit, where it freezes into a layer of icy fog. Pileus clouds are often short-lived, with the main cloud beneath them rising through convection to absorb them. A pileus or cap cloud should not be confused with an incus ('anvil'), the yet more visibly hairy and striated area of ice crystals that develops above a Cumulonimbus capillatus cloud (C_L9).

RIGHT:
Cumulus congestus with pileus (Cu con pil), a frozen cap of moist air lying flat on top of the cloud.

Pannus (pan)

From the Latin for 'cloth' or 'rag', pannus are accessory clouds in the form of dark, ragged shreds of either Stratus fractus (C_L7) or Cumulus fractus (C_L1), that appear below other clouds, typically those of the genera Altostratus, Nimbostratus, Cumulus and Cumulonimbus. Pannus clouds have historically been referred to as 'scud' or 'messenger' clouds, particularly by sailors and farmers, for whom the message that they brought was: rain.

Pannus clouds are often attached to the main body of other, precipitating cloud layers, but they can sometimes form a distinct layer in their own right, obscuring the main cloud above them, as in this photograph, taken on a rainy day in Brynmawr, Gwent, of dense, dark pannus clouds swirling below a whitish layer of Altostratus.

RIGHT:
Stratus fractus pannus (St fra pan) cloud lowering above the town of Brynmawr, Gwent.

Velum (vel)

From the Latin for 'sail' or 'awning', a velum or veil cloud is a thin, wide, long-lasting layer of low altitude cloud through which the summits of Cumulus congestus (C_L2) or Cumulonimbus clouds will often pierce. In contrast to icy pileus clouds (page 110), which are generally short-lived, velum clouds are formed in layers of stable, humid air that are lifted by the convection currents within the main cumuliform clouds, and they can persist even after their host clouds have dispersed or decayed.

RIGHT:
A **Cumulonimbus capillatus** cloud rising through a stable layer of low altitude cloud known as a **velum**, or 'veil'.

Flumen (flm)

Flumen ('flow cloud') are bands of low clouds associated with severe convective Cumulonimbus storm clouds, arranged parallel to the low-level winds and moving into or towards the storm cell. The cloud elements are drawn towards the updraft into the supercell, the base being at about the same height as the updraft base. Note that flumen are not attached to the murus wall cloud (see murus, page 120) and that the cloud base is higher than the wall cloud. There is a variety of inflow band cloud known as the 'beaver's tail', which is distinguished by its broad, paddle-like appearance.

BELOW:
Flumen appearing
near the base of a
Cumulonimbus storm
cloud, with its distinctive
'beaver's tail' formation.

Supplementary Features

Incus (inc)

The anvil-shaped summit of a large Cumulonimbus capillatus cloud (C_L9), the incus (Latin for 'anvil') is an icy canopy that can grow to enormous heights above the main body of the cloud, spreading out laterally when it meets the tropopause – the atmospheric boundary separating the troposphere from the stratosphere (see glossary: *atmosphere*) – to create the charateristic flattened thunderhead seen here. It can be smooth in appearance, especially at a distance, but it is usually highly fibrous and striated, being composed of billions of ice crystals borne aloft by vigorous upward convection.

The incus can sometimes separate itself from the main storm cloud, preceding it by many miles, even producing cloud-to-ground lightning in the apparent absence of the main storm cloud. If, as sometimes happens, the main Cumulonimbus cloud decays or rains itself out, leaving the anvil behind on its own, the resulting cloud is known as Cirrus spissatus cumulonimbogenitus (see C_H3, page 90).

RIGHT:
The flattened, icy thunderhead of the **Cumulonimbus capillatus** (Cb cap inc), rising over the town of Mosas, southern Sweden.

Mamma (mam)

Mamma (also known as 'mammatus') are udder-like protuberances that can form on the under surfaces of Stratocumulus or Cumulonimbus clouds, particularly underneath the anvils of the latter. They are caused by powerful downdraughts, when pockets of cold, moist air sink rapidly from the upper to the lower parts of the cloud, reversing the usual cloud-forming pattern of the upward convection of warm, humid air. Their shapes and forms can vary considerably, from near-spherical pouches to tubular, rippled or merely undulating globules, arranged in cellular formation.

This dramatic image, taken at Brize Norton airfield in Oxfordshire, shows large mamma formed beneath a powerful Cumulonimbus capillatus cloud (see C$_L$9, page 48).

RIGHT:
Undulating **mamma**
forming below
**Cumulonimbus
capillatus** cloud, RAF
Brize Norton, Oxfordshire.

Virga ('fallstreaks')
(vir)

Virga (from the Latin for 'rod') is any form of precipitation, whether rain, snow or ice, that evaporates before it reaches the ground. Its failure to carry on all the way down is usually due to its passing through a layer of warmer or dryer air, although sometimes atmospheric conditions will change, and the virga will be replaced by real precipitation from the same cloud (see page 118).

Often wispy or hooked in appearance, virga is most associated with high- or medium-level cloud formations, as in this photograph of tendrils of virga falling from an Altocumulus floccus (C$_M$8).

RIGHT:
Tendrils of **virga** falling from an **Altocumulus floccus** cloud. Such fallstreaks will usually evaporate long before reaching the ground.

Arcus (arc)

A distinct shelf or roll of low cloud that can
appear below a powerful Cumulonimbus cloud,
an arcus (from the Latin for 'arch') is formed
by strong downdraughts of cold air, which
spread out ahead of the oncoming stormcloud,
pushing up layers of warm air nearer the
ground. These form dense, horizontal rolls
of cloud, some of which (as the photograph
shows) can look quite eerie and menacing.

RIGHT:
**Cumulonimbus
capillatus with arcus**
(Cb cap arc),
Hill City, Kansas.

Praecipitatio
(pra)

From the Latin for 'fall', the term *praecipitatio* is
applied by meteorologists to a cloud from which
any kind of rain, snow or hail manages to reach
the ground, as distinct from virga (see page 116).
Although the two examples shown here are from
Cumulonimbus clouds, precipitation can fall from
a variety of other cloud types, including Stratus,
Stratocumulus, Altostratus and Nimbostratus clouds.

RIGHT:
A hail shower falling
over Kansas. Hail forms
when warm updraughts
of air hurl descending
ice pellets back up into
the colder regions of the
cloud. The pellets grow
through collision and
freezing, creating ever
bigger stones.

OPPOSITE:
A heavy shower falling
from a Cumulonimbus
cloud, Arkansas.

Murus (mur)

Cauda (cau)

Murus ('wall cloud') is another supplementary feature associated with strong Cumulonimbus storm clouds. These features (which are sometimes known as 'pedestal clouds') are isolated lowerings attached to the storm cloud's rain-free base, and indicate an area of strong updraft in which rain-cooled air is pulled towards the storm cloud's core. Murus clouds thus mark the area of strongest updraft within the storm, and are characterized by strong winds; in fact, tornadoes and tuba ('funnel clouds') often form within wall cloud structures. Murus is also sometimes accompanied by a cauda (see next entry).

Cauda ('tail') is a horizontal, tail-shaped cloud (not a funnel, see page 126) that can extend from the main precipitation region of a supercell Cumulonimbus; it is typically attached to the wall cloud (see murus, left), with cloud motion moving away from the precipitation area and towards the murus to which it is attached. Most movement is horizontal, but some rising motion is often apparent as well. Ragged-looking cauda is easily mistaken for the accessory cloud flumen (see page 113): both are types of inflow bands, but the cauda feature is attached to the storm's wall cloud (murus), while the flumen feature can be significantly larger and appear higher up in the storm structure, where it feeds into the storm cloud's updraft.

OPPOSITE:
Murus ('wall cloud'), accompanied by a distinctive **cauda** ('tail') at the base of a large **Cumulonimbus** storm cloud.

Cavum (cav)

Cavum (otherwise known as a 'fallstreak hole' or 'hole-punch cloud') is an effect caused by the sudden freezing of an isolated patch of supercooled cloud, which falls away to leave a visible gap in its place. The result is usually a well-defined circular (though sometimes linear) hole in the high cloud layer, from which virga or wisps of Cirrus can be seen to descend. Cavum is typically a circular feature when viewed from directly beneath, but may appear oval shaped when viewed from a distance. The physical cause of the phenomenon is not yet fully understood, though it seems to happen only in supercooled clouds, in which water droplets remain in liquid form even in subzero temperatures. It is possible that particulates from aircraft exhaust play a part in the creation of cavum, since supercooling often occurs when there are not enough freezing nuclei available for airborne water droplets to turn into ice. Fallstreak holes are often observed in the wake of aircraft interaction, though these will generally be linear, in the form of dissipation trails, with virga falling from the progressively widening trails. Cavum usually occurs in Altocumulus and Cirrocumulus clouds, though has occasionally been observed in Stratocumulus.

RIGHT:
Cavum ('fallstreak hole') in a patch of icy **Cirrocumulus** over New Zealand's South Island.

OPPOSITE:
Cavum, also known as a 'fallstreak hole' or 'hole-punch cloud', appearing in a layer of **Altocumulus stratiformis** above Oklahoma.

Asperitas (asp)

Asperitas (from the Latin for 'roughened) are well-defined, wave-like structures that appear on the undersides of Stratocumulus and Altocumulus clouds. They are more chaotic and show less horizontal organization than the variety undulatus, with localized waves that appear either smooth or dappled with smaller features, sometimes descending into sharp points, as if viewing a patch of roughened sea from below. Asperitas was first identified by members of the Cloud Appreciation Society, and was the first of the twelve new official cloud terms to be adopted by the World Meteorological Organization in 2017.

RIGHT:
A swirling display of
**Altocumulus stratiformis
asperitas** over the city
of Tallin, Estonia.

Fluctus (also known as Kelvin–Helmholtz waves) (flu)

Fluctus (from the Latin for 'wave') is a relatively short-lived formation that usually appears on the upper surface of a Cirrus, Altocumulus, Stratocumulus or Stratus cloud, in the form of curls or breaking waves. They appear when the boundary between a warm air mass and a layer of colder air beneath it is disturbed by strong horizontal winds, causing the upper layer to move faster than the lower. This aerial turbulence, or wind shear, causes the 'crests' of the waves to move ahead of the main body of the cloud, leading to the characteristic wave formation seen here.

This rare, short-lived phenomenon was originally named after two nineteenth-century scientists who pioneered the study of turbulent flow: the Belfast-born physicist William Thomson, 1st Baron Kelvin (1824–1907), and the German physicist Hermann von Helmholtz (1821–94). The new name, 'fluctus', was added to the official cloud classification in 2017.

BELOW:
Fluctus (Kelvin–Helmholtz waves) form in distinct layers of turbulent air moving at different speeds.

Tuba
('funnel cloud') (tub)

A tuba (or 'funnel cloud') sometimes forms at the base of a Cumulonimbus cloud when a column of swirling air begins to rotate, condensing ambient moisture into water droplets. This vortex begins to move downwards, creating a tapered cone or funnel shape that protrudes some distance below the cloud, although it is rarely strong enough to make contact with the ground. When it does, it usually takes the form of a weak landspout or waterspout (see opposite), rather than a fully fledged tornado, which tends to develop from the large-scale rotation of a tropical supercell thunderstorm, in contrast to the weak vorticity of the cold air funnel cloud.

Luke Howard once described a funnel cloud that he spotted near the Yorkshire coast, in a letter dated

RIGHT:
Funnel cloud,
Kansas.

17 July 1851 (noon): 'We have been entertained within this hour past by the appearance of that very rare phenomenon the Water-spout. The clouds were heavy to the West and N. West with Rain behind them; when one pretty much in advance of the shower let down the usual Jelly-bag – as the women very aptly described it – but only about half-way to the earth, and not being over any water, as I conclude, there was no reaction that we could perceive from beneath.' Describing it as a waterspout ('tis Neptune shaking hands with Jove', as he excitedly concluded), seems to have been an extrapolation on his part, as it is clear from his detailed description that what he actually witnessed was a funnel cloud.

RIGHT:
When a cold-air funnel cloud is strong enough to make contact with the ground, it usually does so in the form of a **landspout** or (if over water) a **waterspout**, as in this example seen over Yarmouth, Isle of Wight. Often mistaken for tornadoes, such spouts are much weaker and shorter-lived, rarely causing anything in the way of damage.

Special Clouds

Nacreous ('mother of pearl') cloud

Nacreous clouds, also known as polar stratospheric clouds (PSCs), appear high in the atmosphere, some 15 to 30km (10 to 20 miles) above the Earth, generally in latitudes higher than 50°, particularly in the northern hemisphere. They form in the freezing temperatures of the lower stratosphere, often below -80°C (-112°F), and are usually a mixture of nitric acid and ice crystals, sourced from parcels of moist air that are forced up through the tropopause (see glossary: *atmosphere*) by the same orographic oscillations that are responsible for producing high lenticular wave clouds (see C_M4).

The likeliest time to see them is during a winter sunrise or sunset, when most of the sky is dark, leaving them lit by the Sun from beneath the horizon.

Their iridescent pastel colours can be magically beautiful, an effect heightened by their enormous distance from the viewer.

But there is a dark side to these iridescent wave clouds: their chemical composition assists the production of chlorine atoms, which in turn contributes to the depletion of the ozone layer, 25km (15 miles) above the Earth, where nacreous clouds are mostly found. A single chlorine atom can destroy up to 100,000 ozone molecules — which is why the large-scale release of chlorofluorocarbons (CFCs) into the atmosphere was to prove such a disaster — so, sadly, these most beautiful and benign-looking clouds turn out to have a powerfully destructive environmental impact.

OPPOSITE:
The iridescent pastel colours of a **nacreous cloud**, photographed in the night sky above North Yorkshire (54° north).

Noctilucent clouds

Noctilucent clouds (NLCs), also known as polar mesospheric clouds, are the highest clouds in the Earth's atmosphere, occurring in the mesosphere (the layer immediately above the stratosphere) at altitudes above 80km (50 miles), at least four times higher than those of any other cloud in this book (with the single exception of stratospheric nacreous clouds). Being so high, they catch the last light of the setting Sun long after the Earth's lower atmosphere has been plunged into shadow, hence their name, which derives from the Latin for 'night-shining'.

NLCs are rarely seen, although they seem to have become less rare over the past 30 years, which may or may not be due to human activity (see 'Clouds and Climate Change', page 155). The best chance of spotting one is on a clear midsummer night somewhere between 50° and 65° latitude (north or south), such as northern Scotland or Scandinavia (as in the example shown here), when they are underlit by the last of the Sun.

Appearing as thin, milky-blue or silvery waves high in the sky, on the fringes of space, NLCs might appear to be random in their pattern structure, but there are in fact four main types: veils, bands, billows, and whirls. These types rarely appear as single displays, however, but tend to mix and merge in the sky together. Noctilucent clouds remain the least understood clouds of all, the mechanics of their formation in such dry, clear, intensely cold atmospheric conditions (-125°C/-193°F) having not yet been established, although many hypotheses have been advanced, including the idea that they seed themselves from meteorite debris, or from dust blasted high into the stratosphere by major volcanic eruptions on Earth, or even from the constituent elements of space shuttle exhaust fumes. The meteorite dust hypothesis is currently the most favoured, and has been strengthened by a discovery made by the solar-powered spacecraft Mars Express, that clouds of CO_2 with a similar appearance to NLCs are to be found high in the atmosphere above the Red Planet: Earth's distant noctilucent neighbours.

OPPOSITE:
The least understood cloud of all, **noctilucent clouds** appear as thin, milky waves of colour high in the sky on midsummer nights.

Banner cloud

Banner clouds, such as this impressive example, are caused by the physical presence of the mountain itself, which acts as an obstacle to the moisture-laden westerly wind, forcing it up towards the cloud-forming layer, just above the mountain's peak. Once this orographic cloud has been formed (orographic clouds are those created or influenced by mountains), it is pulled down the lee side by the reduced pressure on that side of the mountain, a motion that leads to the cloud's distinctive streaming pennant formation. The Matterhorn's banner clouds are the best-known examples of this phenomenon, although other large peaks, such as the rock of Gibraltar, create their own outstanding examples.

RIGHT:
Banner cloud streaming westwards from the east face of the Matterhorn, Switzerland.

Cataractagenitus

Clouds can develop in the vicinity of large waterfalls, from falling water broken up into ambient moisture. The plummeting cascade of water drags air down with it, causing neighbouring regions of moist air to lift up and replace it. This movement creates favourable cloud-forming conditions above the falls. Cataractagenitus clouds (the term derives from the Latin *cataracta*: 'waterfall'), which tend to be small in size, are given the name of the appropriate genus, followed by species or variety, plus the special cloud name, for example: Cumulus humilis cataractagenitus or Stratus fractus cataractagenitus.

RIGHT:
A summertime display of **Stratus fractus cataractagenitus**, generated by the cascading waters of Niagara Falls.

Silvagenitus

Clouds often develop locally over forested areas as a result of increased humidity due to evaporation and evapotranspiration from the tree canopy. The process of forest cloud formation can often be seen on sunny mornings after rain, with spectral fingers of cloud appearing to rise and gather above the trees. Where these special clouds are observed, they are given the name of the appropriate genus and any appropriate species or variety, along with the special cloud name silvagenitus (from the Latin *silva*: 'forest'), as in this early morning example of Stratus silvagenitus, opposite.

RIGHT:
Stratus silvagenitus forming over the western slopes of the Andes, north of Quito, Ecuador.

OPPOSITE:
Wisps of **Stratus silvagenitus** appear over an area of forest after a bout of summer rain.

Man-made Clouds

Flammagenitus

Clouds may develop as a consequence of convection initiated by heat from forest fires, wildfires or volcanic eruption activity. Such naturally occuring clouds are denoted by the special cloud term flammagenitus (from the Latin for 'fire-made'), as, for instance, Cumulus congestus flammagenitus, or Cumulonimbus calvus flammagenitus, the latter being large clouds that appear above erupting volcanoes, or extensive areas of wildfire.

Cumulus flammagenitus clouds have long been known by the unofficial name 'pyrocumulus', a hybrid Greek–Latin term that does not distinguish between clouds produced by natural or human activity, such as stubble-burning or other forms of ground-level combustion, as in the case of this low-lying, smoke-tinged Cumulus humilis flammagenitus cloud hovering eerily over a field of burning stubble in Wiltshire.

The air needs to be fairly still for viable flammagenitus clouds to form, otherwise the thermal currents cooked up by the fire will be dispersed before the rising water vapour can reach the condensation point necessary for the formation of clouds. Once the fire is out, the cloud will soon decay and disappear, as would any other small cumuliform cloud when the convective currents that created it subside.

RIGHT:
Cooked up from the thermal currents above a burning field of stubble near Salisbury, a **pyrocumulus** (Cumulus humilis flammagenitus) cloud hovers eerily in the evening sky.

Cumulus homogenitus ('fumulus')

A variety of man-made Cumulus cloud, 'fumulus' clouds typically form above industrial cooling towers. Much of the moisture that rises and condenses to form such clouds is emitted from the towers themselves over sustained periods, and this can combine with moisture already present in the atmosphere to produce significant and long-lasting cloud formations, as seen here. These clouds are denoted by genus and species, followed by the special cloud name homogenitus (from the Latin for 'man-made'), in this case, Cumulus mediocris homogenitus.

RIGHT:
Cumulus mediocris homogenitus ('fumulus') cloud formed above the cooling towers of the now-demolished Didcot Power Station, Oxfordshire.

Condensation trails ('contrails')

The only cloud type that Luke Howard could never have witnessed, contrails (also known as 'vapour trails') are formed by the sudden condensation of the water vapour ejected from the exhaust of a jet aeroplane. At altitudes of 11km (35,000ft) or more, the outside temperature is well below freezing, so most of the contrails that we see are, like their neighbouring natural Cirrus clouds, formed entirely of slowly sinking ice crystals (see the photograph opposite for an example of a contrail breaking up, or rather breaking down, high in the sky). For contrails to appear and persist, the air outside the aircraft must be cold and moist. If the air is too dry, or not sufficiently cold, no contrails will appear at all, or a very short-lived water droplet variety will briefly trail the plane before evaporating into the surrounding air. If a contrail persists for ten minutes or more, it is classified as a special cloud, Cirrus homogenitus (from the Latin for 'man-made'). In the past, contrails were sometimes referred to as 'Cirrus aviaticus', but in

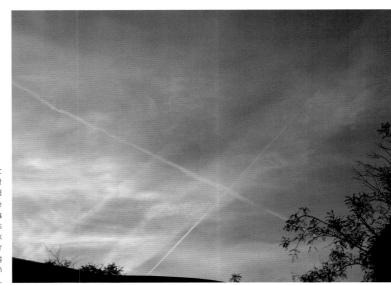

RIGHT:
Persistent aircraft
contrails that spread
across the sky are
known as **Cirrus
homomutatus**. This
image shows a mix
of new and older
contrails covering
the sky near London
City airport.

2017 the term 'homogenitus' was formally adopted for international use.

In particularly moist upper air conditions, Cirrus homogenitus clouds can persist for long periods of time and, under the influence of upper winds, spread for many kilometres across the sky. When this occurs, the clouds are denoted by their relevant species and variety, followed by 'homomutatus' (for example, Cirrus spissatus homomutatus).

The opposite of a contrail, known as a distrail (short for 'dissipation trail'), occurs when an aircraft flies through a natural Cirrus or Altostratus cloud, its vapour trail serving to overload the cloud with extra-heavy supercooled water droplets (or ice crystals), which then fall out, leaving a linear gap in its wake. See also cavum, page 122.

Contrails may be unpopular with some skywatchers, who regard them as a form of visual pollution, but their behaviour can offer useful clues as to what is going on in the upper atmosphere. In temperate latitudes, for example, when a contrail is seen to persist, spreading out over many hours into Cirrus homomutatus, this is a sign that moist air in the upper atmosphere is rising, meaning that rain is likely to arrive over the next 48–72 hours. The absence of contrails (in temperate latitudes) suggests, by contrast, that fine, dry weather is likely to continue over the same period of time. Contrails may also have longer term significance, as factors in global climate change, as is discussed in the Afterword on page 155.

BELOW:
Ice falling from an aircraft **contrail** (Cirrus homogenitus), Berkshire.

Optical Phenomena and Effects

Halo

A halo is an optical phenomenon that appears around the Sun or Moon when differently angled ice crystals in high Cirrostratus or Cirrus clouds reflect and refract the incoming light, splitting it into faint rainbow colours. Haloes are usually 22° in radius, and thus much larger than a typical corona (see page 144). Haloes often appear in conjunction with other optical effects, such as Sun dogs and parhelic circles (see opposite), and are often indicative of bad weather to come, as Cirrostratus clouds tend to spread out ahead of developing and advancing frontal systems – 'haloes around the Moon or Sun means that rain will surely come', as an old weather saying succinctly puts it.

Haloes can also occur around other powerful light sources, such as streetlamps, as well as (in unusually cold conditions) above collections of minute ice crystals at ground level, known as 'diamond dust'.

RIGHT:
A **22° halo** around a streetlight in Newquay, Cornwall, formed by the refraction of light through atmospheric ice crystals.

Mock sun / Sun dog / parhelion

Appearing as bright spots on either side of the Sun, mock suns (also known as Sun dogs, or parhelia, from the Greek meaning 'with the Sun') are another form of halo phenomenon, caused by the refraction of sunlight by ice crystals present in high cirriform clouds, especially Cirrostratus. They tend to occur when the Sun is low in the sky, and will appear to move closer to the Sun the lower it sinks. Parhelia often exhibit a range of rainbow colours, with the red end of the spectrum appearing towards the inner edges, nearest the Sun, and the bluish end appearing towards the outer ones.

It is the particular shape and orientation of the ice crystals within the cloud that determine whether sunlight will be refracted as a halo or a parhelion. Flat, horizontally aligned crystals will tend to produce parhelia, whereas more randomly shaped or poorly aligned crystals will tend to produce the 22° halo (see previous entry).

Moon dogs and lunar haloes are much rarer versions of the same effects.

RIGHT:
A **Sun dog** appears alongside evening clouds, an effect caused by the refraction of sunlight by ice crystals high in the sky.

Sun pillar

A vertical streak or blade of light that appears above or below a setting or rising Sun, sun pillars are produced by light reflected from the horizontal surfaces of slowly falling, hexagonally shaped ice crystals that are present in cirriform clouds, especially Cirrostratus clouds (see C_H7, page 100). Unlike refraction phenomena such as haloes and sun dogs, pillars are simply the collective glittering of millions of tiny icy crystals, and will thus reflect the colours of the surrounding sunset, rather than divide into the spectrum colours that are associated with the process of refraction.

On rare occasions, a sun pillar will coincide with a parhelic arc, forming a bright cross in the sky, centred on the Sun.

RIGHT:
A **sun pillar** in the early evening sky over Sandhurst, Surrey.

Circumzenithal arc

Caused by the refraction of sunlight through the same horizontally aligned cirriform ice crystals that are responsible for producing Sun dogs (see page 141), a circumzenithal arc (CZA) is a band of bright prismatic colours that resembles an inverted rainbow, positioned immediately above the viewer's head (i.e., at the zenith, the point in the sky immediately above any given position).

Typically a quarter-circle in shape, its colours, which are often brighter than those of the rainbow, run from blue near the zenith, down to red near the horizon. Due to the precise angle in which the refracted light exits the sides of the horizontal ice crystals, CZAs cannot occur if the Sun is more than 32.2° above the horizon, and the brightest arcs of all – which the cloud writer Gavin Pretor-Pinney has archly dubbed 'the cloud smile' – occur when the Sun is exactly 22° above the horizon.

RIGHT:
An ice rainbow, known as a **circumzenithal arc**, makes an appearance in the early evening sky.

Corona

A corona is a coloured ring or sequence of rings surrounding the Sun or Moon, as seen through thin layers of low or medium cloud. As can be seen in this impressive example of a corona in Altocumulus, the colours of the corona tend to be bluish or whitish nearest the centre, shading towards the red end of the spectrum further out. As coronas are caused principally by the diffraction (bending and scattering) of light by uniformly sized water droplets, the phenomenon is closely related to that of irisation (see opposite).

A less common form of the corona is the Bishop's ring effect, caused by dust and sulphate particles lingering in the atmosphere following major volcanic eruptions. The effect (named after Sereno Bishop, who first described the phenomenon in the wake of the eruption of Krakatau in August 1883), is of a wide corona around the Sun, the inner rim of which is whitish or bluish in colour, while the outer part tends to reddish, and even purplish. The effect was widely documented in 1991–92, following the eruption of Mount Pinatubo, in the Philippines, in June 1991, although sightings were also reported across Northern Europe in April and May 2010, following the eruption of the Icelandic volcano Eyjafjallajökull.

RIGHT:
Corona appearing in Altocumulus clouds, Duntulm, Isle of Skye.

Irisation

Irisation is a form of light interference caused by uniformly sized water droplets serving to diffract sunlight, bending the light round the drops rather than allowing it to pass straight through. The light typically recombines to form irregular patches of luminous pastel or mother-of-pearl shades, predominantly green and pink, at the edges of thin Cirrocumulus, Altocumulus and Stratocumulus clouds. The word is derived from Iris, the Greek goddess of the rainbow.

RIGHT:
Irisation in Altocumulus clouds, The Needles, Isle of Wight.

145

Glory

A glory is an optical phenomenon produced by light scattered back towards its source by a cloud of uniformly sized water droplets. Appearing as a sequence of coloured rings, glories are sometimes seen by mountaineers in association with the so-called Brocken spectre, the magnified shadow of a person that is cast by low sunlight on to the upper surfaces of clouds that are situated below where the viewer is standing. The Brocken spectre is named after the cloud-shrouded peak in the Harz mountains of northern Germany, where the effect was first recorded.

RIGHT:
Glories are now most often observed from aeroplanes, as in this example, taken from the window of a plane passing over the Whitsunday Islands off the coast of Queensland, Australia.

OPPOSITE:
This haunting example of a **glory** accompanying a **Brocken spectre** was taken on a misty mountainside; note how the glory's concentric rings are centred on the photographer's own head. Two mountaineers, standing side by side, would each see his or her own, individual glory, and not that of his or her companion.

Rainbow

Rainbows are caused by water droplets dispersing sunlight into banded arcs made up of the seven colours of the visible spectrum: red, orange, yellow, green, blue, indigo, and violet. White light is refracted as it enters the droplet, then reflected off the back surface, and rerefracted on the way out, being thereby dispersed in a wide range of angles, the most visually intense of which will always be between 40° and 42°, due to the spherical nature of the droplets. Rainbows can only be seen when the Sun is behind the observer, and the airborne water droplets — whether from falling rain, spray from a waterfall, or even from a garden sprinkler — are directly in front. Being entirely optical phenomena (rainbows do not exist in any tangible sense), the eye of every observer will constitute its own individual rainbow, with a unique position in the sky that is always exactly opposite to the Sun with respect to where each individual observer happens to be standing. All rainbows are technically circular, but when seen from the ground the lower half tends to dip below the horizon and is thereby rendered invisible. Complete circular rainbows can often be seen from higher up in the sky, from the vantage point of an aircraft window.

The most commonly seen bow is the primary bow, which is always 42° in radius, with red on the outer edge and violet on the inner edge. Sometimes a larger secondary bow (52° radius) accompanies the first, as can just be seen in the top of the photograph opposite, in which the colours are reversed, red on the inside, violet on the outside. Pale arcs, known as supernumerary bows, also sometimes appear inside the primary bow.

LEFT:
Primary rainbow
Hexham,
Northumberland.

RIGHT:
Primary and secondary rainbow Embleton Bay,
Northumberland.

Crepuscular rays

Crepuscular rays are beams of sunlight that are scattered and made visible by minute particles within the lower atmosphere, especially dust, gases and water droplets. There are three recognized varieties of ray, their differences due to variations in atmospheric transparency – 1: rays which emerge from gaps in low cloud, caused by the scattering of light by water droplets; 2: beams of sunlight which rise from behind a cloud (usually a cumuliform cloud); and 3: pinkish rays which radiate from below the horizon, their intense coloration caused by the presence of haze. The downward streaming 'Jacob's Ladder' variety, as seen here, takes the form of columns of sunlit air, separated by shadows cast by the intervening cloud, which appear to converge on the Sun. The effect takes its name from the episode in Genesis 28: 11–17, in which Jacob dreams of a ladder connecting earth to heaven, on which hosts of angels could be seen ascending and descending.

A rare counterpoint to crepuscular rays are anti-crepuscular rays, which occur in exactly the same way as the former, but instead of appearing to converge towards the Sun, they appear to be scattering away from it, due to a simple trick of perspective that occurs only when the viewer's back is turned against the setting (or recently set) Sun, and he or she is looking towards a point on the opposite horizon, known as the antisolar point. The poet Gerard Manley Hopkins, a life-long sky-watcher, wrote a memorable diary description of upward-pointing crepuscular rays on the stormy afternoon of 30 June 1866: 'Thunderstorms all day, great claps and lightning running up and down. When it was bright betweentimes great towering clouds, behind which the Sun put out his shaded horns very clearly and a longish way.'

LEFT:
Crepuscular rays,
'Jacob's Ladder' variety
Jersey, Channel Islands.

Aurora

Appearing as tall ribbons of multicoloured light, or as more diffuse glowing patches in the sky, *c*.100 to 250km (60 to 150 miles) above polar latitudes, aurorae are produced by fast-moving particles from the streaming solar wind as they collide with atmospheric gases in the Earth's upper atmosphere, causing gas molecules to fluoresce in a range of different colours. Aurorae can be seen at both poles, and are known as *aurora borealis* (or 'The Northern Lights') in the north, and as *aurora australis* in the south. They are named after Aurora, the Roman goddess of the dawn.

RIGHT:
The Northern Lights:
formed as the solar wind
collides with Earth's
gaseous atmosphere.

Lightning

A visible electrical discharge produced by the actions of Cumulonimbus thunderclouds, lightning is formed when pockets of positive (+) and negative (-) charges within an individual cloud become separated by the action of violent updraughts of air. Clusters of electrical charge will accumulate in different parts of the cloud, with positive charges often congregating at the top, and negative charges at the base of the cloud. When the potential difference between these charged areas becomes too great to sustain, electrical energy is discharged in the form of lightning.

Lightning strikes the ground when the negative charges in the base of the cloud induce opposite charges in the ground below, sending a powerful spark flying between the differently charged regions in the form of cloud-to-ground lightning (often described as forked lightning). Sparks that fly between differently charged regions within an individual cloud, and thus do not reach the ground, are known as in-cloud (or sheet) lightning, while sparks that fly across the air from one charged cloud to another are known as cloud-to-cloud lightning.

What appears to be a single bolt of lightning usually takes the form of a sequence of near-instantaneous discharges, which seek the ground in a series of 'steps'. Thunder is the sound created by the sudden expansion of the air column as the lightning surges through at temperatures of up to 28,000°C (50,430°F). Whether one hears a sudden crack, or a low rumbling sound depends upon how short the stroke was, as well as how far away one is from the scene of the strike. Thunder can sometimes be heard nearly 30km (20 miles) away, but over those kinds of distances the sound-waves will have broken up into vague, indistinct rumbles.

LEFT:
Cloud-to-ground lightning, Lubbock, Texas.

OPPOSITE:
Lightning often seeks the ground in 'steps', as can be seen in this dramatic photograph of forked lightning striking the Earth near a road in Archer County, Texas.

AFTERWORD: CLOUDS AND CLIMATE CHANGE

As will have been apparent during the course of this book, clouds often play a valuable role in indicating short-range weather conditions, but when it comes to predicting longer term climatic changes, they remain largely unknown quantities. For despite the near-universal scientific consensus on the reality of global climate change, the subject remains riddled with uncertainties, among the most pressing of which concerns the likely role that clouds will play in shaping future conditions on Earth. Will clouds turn out to be agents of global heating, serving to veil us in an ever-thickening blanket of greenhouse emissions, or will they help save the day by reflecting ever more sunlight back into space? These, it turns out, are far from simple questions, and as the most recent (2013–14) published assessment report by the Intergovernmental Panel on Climate Change made clear, 'clouds and aerosols continue to contribute the largest uncertainty to estimates and interpretations of the Earth's changing energy budget'. [1] A change in almost any aspect of clouds, such as their type, location, longevity, water content, altitude, particle size or shape, will affect the degree to which those

clouds either warm or cool the Earth. Some changes amplify warming while others diminish it, and a great deal of current research is directed towards a better understanding of how clouds change in response to atmospheric warming, and how these changes might in turn affect climate through a variety of complex feedback mechanisms.

As is so often the case with climate science, much of this ongoing research yields potentially contradictory results. On the one hand, for example, many climate scientists believe that continued surface warming will see an increase in water vapour rising from the oceans, thus leading to an overall increase in cloud formation; while on the other, it has been suggested that, in warmer latitudes, an increase in atmospheric water vapour content would see large convective cumuliform clouds building up and raining themselves out far quicker than they do at present, thereby leading to a net decrease in the Earth's total cloud cover. It is not certain which outcome is the more likely, nor do climate scientists fully understand the kind of long-term influences that either would be likely to have. Even if, for the sake

of argument, we assume that overall cloud cover will increase as the surface of our planet continues to warm, it remains unclear what kind of clouds (and thus what kind of feedback scenarios) are likely to predominate. For instance, high, thin cirriform clouds, such as Cirrostratus, tend to have an overall warming effect, as they admit a lot of shortwave radiation in from above (in the form of sunlight during the day), while intercepting longwave back-radiation (warmth reflected from the sunlit ground) and despatching it

OPPOSITE:
Low, thick cloud cover cools the Earth by reflecting sunlight back into space.

back down to Earth. [2] Any increase in Cirrostratus cloud cover would therefore result in the addition of yet another warming mechanism to our climate. In contrast, however, bright, dense clouds, such as Cumulus congestus, serve to cool the Earth by reflecting incoming sunlight back into space by day. At night, these same clouds can exert a slight warming effect, by absorbing or reflecting back-radiation, but their overall influence is a cooling one, especially when their summits grow reflectively dense and white. So, in theory, an increase in high, thin layers of cloud would amplify the warming effect, while an increase in low, dense, puffy clouds would exert a contrary cooling influence. In reality, of course, things are never so simple, and, as has been seen throughout this book, clouds have a habit of behaving in unpredictable ways. By way of illustration, here are just two examples of such cloudy complexity in action:

The first occurred in the wake of the terrorist attacks of September 11, 2001, when all commercial flights in the United States were grounded for several days, leaving the skies contrail-free for the first time since the 1920s. The result, according to a comparison of nationwide temperature records, was slightly warmer days and slightly cooler nights than were usual for that time of year, the normal night/day temperature range having increased by 1.1°C (34°F). According to the climate scientists who worked on the data, this was a likely consequence of additional sunlight reaching the surface by day, and additional radiation escaping at night through the newly contrail-free skies. [3] At first sight this might appear counter-intuitive, for surely the Cirrus homomutatus clouds created by the spreading of aircraft contrails are straightforward warming clouds, the kind that admit sunlight in from above, while redirecting back-radiation down to the surface. An absence of contrails ought, therefore, to have had an overall cooling effect. But contrails are a lot more complicated than that. When they are in their initial, water droplet, stage they are far denser than natural Cirrus clouds, since they are created from two distinct sources of vapour: the hot moisture emitted by the aircraft's exhaust plus the cold moisture already in the atmosphere, condensed into a turbulent mixture of water droplets and ice crystals, which seed themselves on the solid particulates present in the exhaust plume (see glossary: *condensation nuclei*). At first, this young, opaque contrail (Cirrus homogenitus) behaves more like a white, low-level cloud, reflecting sunlight back into space, and exerting a short-term localized cooling effect. But as persistent contrails begin to spread, they thin out into recognizably cirriform cloud layers, known as Cirrus homomutatus, which often invade

OPPOSITE:
Contrails can at first have a cooling effect on the Earth but as they spread to form layers of **cirriform cloud**, their overall effect becomes warming.

the sky through natural means, their supercooled water droplets having frozen into the kind of tiny icy crystals associated with Cirrostratus formation. Their overall effect thus reverts to a warming one, consistent with the observed behaviour of natural cirriform clouds.

The picture is complicated further, however, by the time of day or night that the contrails form and spread. If contrails spread during the early morning or late evening, they can exercise a slight cooling effect, due to the angle at which sunlight tends to reflect off the ice crystals rather than penetrate through to the ground. At night, by contrast, all clouds, including contrails, can only exert a warming effect, since there is no incoming sunlight to reflect back into space. Any increase in night flights is therefore likely to lead to slightly raised temperatures on the ground. In fact, the warming effects of the predicted increase in contrail production, particularly those associated with a rise in night flights, have been projected to be in the region of a 0.2°C to 0.3°C (32.4°F to 32.5°F) hike per decade in the United States alone (a figure that does not include other warming effects associated with

increased aviation, such as carbon dioxide emissions, soot pollution and local ozone formation). [4]

Of course, much about contrail science remains new and uncertain, and little about these human-made clouds is understood entirely, especially when it comes to the skies above the developing world, where flight continues to grow as a means of transportation, in spite of international agreements to reduce warming emissions. Whether aircraft of the future will need to change the altitudes at which they fly in order to lessen or modify contrail formation, and how the fuel they use could be cleaned, thus containing fewer contrail-generating particulates, is a matter of current research. [5]

Our second example of cloud complexity involves current research into the changing patterns of noctilucent clouds (NLCs) that have become apparent over the past three decades (see page 130). First observed and named in the 1880s, NLCs – also known as polar mesospheric clouds – were once the rarest clouds of all, but not only do they now seem to be a lot more common, they also shine much brighter than they did before, and are observable

from increasingly lower latitudes, with sightings of these blue-tinted clouds advancing steadily towards the equator. [6] Extreme degrees of cold are needed to form such icy clouds in an environment as dry as the mesosphere, 50 to 80km (30 to 50 miles) above the Earth's surface, where temperatures as low as -130°C (-202°F) are normal, and where the air is around 100,000 times dryer than that of the Sahara. According to one now debunked hypothesis, NLCs form around plumes of space shuttle exhaust

OPPOSITE:
A network of **contrails** criss-crossing the English Channel captured by NASA's Aqua and Terra satellites in December 2003.

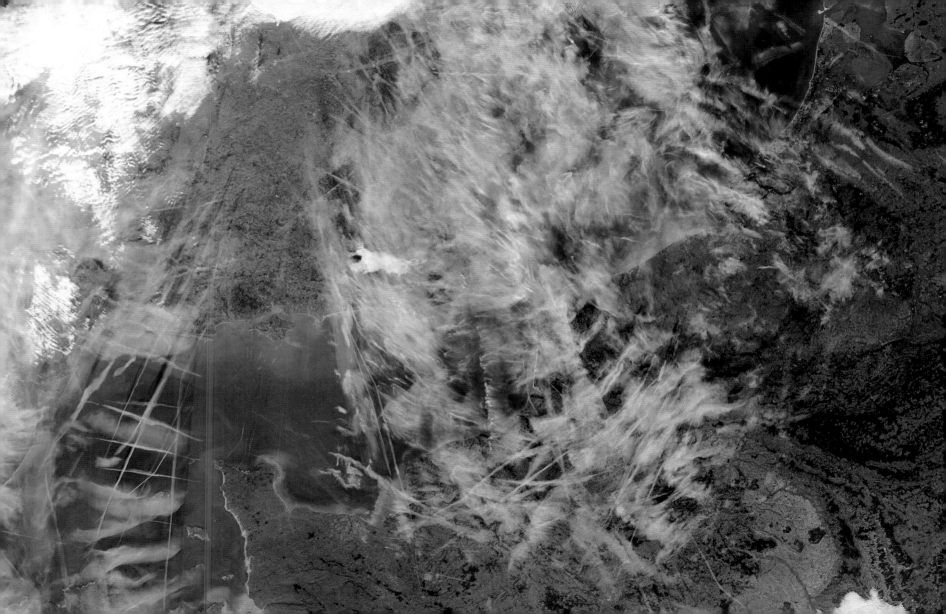

emitted high in the Earth's upper atmosphere, where neither water vapour nor dust nuclei are common natural occurrences, and that therefore these clouds' increased appearance (up by 8 percent a decade) is due to a proportionate increase in space shuttle traffic. More recent research, however, has pointed to ice crystals forming around natural meteoric smoke, nanometre-scale particles that are the remnants of meteors that burned up in our atmosphere.

Strange as it may seem, the increased concentrations of atmospheric greenhouse gases that have contributed to raising temperatures near the Earth's surface have also served to create colder conditions in the Earth's outer atmosphere. This is because greenhouse gases trap much of the longwave surface radiation that has started its return journey out into space; with less thermal energy able to escape from the gas-warmed lower atmosphere, the upper atmosphere is growing correspondingly colder. So the observed increase in noctilucent cloud formation may well be due in part to mesospheric cooling, the lesser known counterpart to global surface warming; so might their increased brightness

be due to more or larger ice crystals forming from a high-altitude influx of water vapour from the warming layers below? After all, NLCs have only been in evidence since the 1880s, the heyday of the Industrial Revolution, so it now seems likely that their widely observed increase will turn out to have an anthropogenic cause. If so, the tangible impact of human activity extends much further into the Earth's upper atmosphere than we could ever have previously imagined. The AIM (Aeronomy of Ice in the Mesosphere) satellite, which was launched by NASA in 2007 on a mission to study noctilucent clouds at close range, has been responsible for many of these findings in the decade and a half since its launch, and as its mission has been extended to include more general environmental data on the Earth's upper atmosphere, it is likely to uncover yet more mesospheric secrets in the future. [7]

'Clouds always tell a true story', as the Victorian meteorologist Ralph Abercromby wrote in 1887, 'but one that is difficult to read.' And though Abercromby was referring to the problem of understanding the relationship between clouds and weather, his

comment can just as aptly be applied to the ongoing exploration of the relationship between clouds and climate. As has already been suggested, there are many interconnected factors which make the life stories of clouds so difficult to read, among the most daunting of which is the fact that, as our climate warms, the atmosphere correspondingly reorganizes itself in ways that could either amplify or mitigate the original warming. Our changing climate has the capacity to alter the day-to-day

OPPOSITE:
First observed in the 1880s, **noctilucent clouds** are now more common, seem to shine brighter and are observable at lower altitudes.

behaviour of clouds and weather in all kinds of unpredictable ways. According to NASA's Earth Observatory, nearly 70 percent of our planet's surface is covered in cloud most of the time, but whether future warming will increase or decrease that percentage is as yet unknown. As the Fifth Assessment Report of the Intergovernmental Panel on Climate Change concluded, 'clouds strongly affect the current climate, but observations alone cannot yet tell us how they will affect a future, warmer climate'. [8] Such is the uncertainty over the likely influence of global feedback mechanisms, especially those involving clouds, the only thing that can be said with any certainty is that clouds could amplify future warming, mitigate future warming, or exert an effect somewhere in between — which, of course, includes having little or no effect at all. In short, we have no way of knowing in advance what is actually going to happen to our planet, and given that, for centuries past, clouds have been metaphors of doubt and uncertainty, it seems as if they are likely to remain so for centuries to come.

[1] *Climate Change 2013: The Physical Science Basis: Contribution of Working Group I to the Fifth Assessment Report of the Intergovernmental Panel on Climate Change* (Cambridge, 2013), p. 573 (via: https://www.ipcc.ch/site/assets/uploads/2018/02/WG1AR5_Chapter07_FINAL-1.pdf).

[2] See ibid., pp. 593–94.

[3] David J. Travis *et al*, 'Contrails reduce daily temperature range', *Nature* 418 (2002), p. 601.

[4] Patrick Minnis *et al*, 'Contrails, Cirrus Trends, and Climate', *Journal of Climate* 17:8 (2004), pp. 1671–85; Lisa Bock and Ulrike Burkhardt, 'Contrail cirrus radiative forcing for future air traffic', *Atmospheric Chemistry and Physics* 19 (2019), pp. 8163–74.

[5] Ulrike Burkhardt *et al*, 'Mitigating the contrail cirrus climate impact by reducing aircraft soot number emissions', *Climate and Atmospheric Science* 1: 37 (2018), via: https://doi.org/10.1038/s41612-018-0046-4.

[6] See James M. Russell *et al*, 'Analysis of northern midlatitude noctilucent cloud occurrences using satellite data and modeling', *Journal of Geophysical Research: Atmospheres* 119: 6 (2014), via: https://doi.org/10.1002/2013JD021017.

[7] Sarah Frazier, 'Taking AIM at Night-Shining Clouds: 10 Years, 10 Science Highlights', NASA Goddard Space Flight Center blog (25 April 2017), via: https://www.nasa.gov/feature/goddard/2017/taking-aim-at-night-shining-clouds-10-years-10-science-highlights.

[8] *Climate Change 2013: The Physical Science Basis*, p. 593.

OPPOSITE:
A mix of **Cirrus homogenitus** and **Cirrus homomutatus** contrails form a densely anthropogenic sky over the riverside city of Nijmegen, Netherlands.

Glossary

accessory cloud

A supplementary cloud which occurs only in conjunction with one of the principal types; they are pannus ('scud'); pileus ('cap cloud'), velum ('veil cloud'), and flumen ('inflow cloud') see pp. 110–113. There are also eleven supplementary features which occasionally appear in connection with the principal clouds: incus ('anvil'), mamma ('udders'), virga ('fallstreaks'), cavum ('fallstreak hole'), praecipitatio ('precipitation'), arcus ('arch'), murus ('wall'), cauda ('tail'), asperitas ('roughness'), fluctus ('wave cloud') and tuba ('funnel cloud'), see pp. 114–127.

Altocumulus (Ac)

A compound of the Latin words *altum*, meaning ('height'), and *cumulus* ('heap' or 'pile'), Altocumulus are medium-level clouds, occurring as individual rounded masses, often with clear sky visible between them.

Altostratus (As)

A compound of the Latin words *altum* ('height') and *stratus* ('layer'), Altostratus are medium-level, dull white or bluish layers of cloud, which tend not to produce rain.

arcus (arc)

From the Latin for 'arch' or 'bow', an *arcus* is an accessory cloud in the form of an arch (see p. 117).

asperitas (asp)

From the Latin for 'roughness', *asperitas* is a wave-like supplementary feature that appears on the underside of Stratocumulus and Altocumulus clouds.

atmosphere

A c. 500-km (c. 300-mile) band of gases encircling the Earth, the atmosphere (from the Greek *atmos* meaning 'air' or 'vapour') is made up of five layers, each with its own distinct temperature profile. 1: the troposphere, extending from the surface to around 20km (c. 12 miles) above the Earth, where most clouds and weather occur, and in which temperature decreases at an average rate of 6.5°C per kilometre of ascent, falling to a minimum of -60°C (-76°F) by the time it reaches the tropopause, the boundary between it and layer 2: the stratosphere, which extends to around c. 50km (c. 30 miles) above the Earth, and in which temperature begins to increase with height, reaching 0°C at its summit, due mostly to the presence of ozone; 3: next comes the mesosphere, extending to c. 85km or more (c. 50 miles) above the Earth, in which the temperature again begins to fall with height, dropping to around -125°C (-193°F) at its summit, where it meets layer 4: the thermosphere, which extends to around 650 km (400 miles) above the Earth, and in which

temperature again rises with height, reaching temperatures of more than 1,000°C (1,832°F) at its summit, where it meets 5: the exosphere, a deep layer of free-moving particles that eventually merges with the emptiness of space.

calvus (cal)

From the Latin for 'bald', the tops of *calvus* clouds appear smooth or bald, in contrast to the hairy or striated appearance of *capillatus* clouds (see C_L3).

capillatus (cap)

From the Latin meaning 'hairy', the term refers to a distinct striated anvil of ice crystals that appears only above Cumulonimbus capillatus clouds (see C_L9).

castellanus (cas)

From the Latin meaning 'castle-like', clouds of the species *castellanus* exhibit cauliflower-shaped turrets at their summits.

cataractagenitus (cagen)

From the Latin *cataracta* ('waterfall') this special cloud is formed in the spray above large waterfalls. Informally known as a 'Niagara cloud' for many years; the term *cataractagenitus* was added to the official cloud classification in 2017.

cauda (cau)

From the Latin for 'tail', this horizontal, tail-shaped cloud is a supplementary feature that extends from the main precipitation region of a Cumulonimbus storm cloud.

cavum (cav)

From the Latin for 'cavity' or 'hole', this supplementary feature, caused by a patch of supercooled cloud freezing and falling away, is also known as a 'fallstreak hole' or 'hole-punch cloud'.

Cirrocumulus (Cc)

A compound of the Latin words *cirrus* ('hair') and *cumulus* ('heap'), Cirrocumulus clouds are thin, white patches or layers of high cloud. A large-scale display of Cirrocumulus is known as a 'mackerel sky'.

Cirrostratus (Cs)

A compound of the Latin words *cirrus* ('hair') and *stratus* ('layer'), Cirrostratus clouds are generally transparent, whitish veils of smooth or fibrous cloud, made up of layers of ice crystals high in the sky.

Cirrus (Ci)

From the Latin for 'hair' or 'fibre', Cirrus clouds are high, white, wispy clouds, often with a fibrous or silky sheen. They are composed of millions of slowly falling ice crystals.

cloud

A cloud is a collection of minute particles of liquid water or ice, suspended in the air and usually not touching the ground. It may also contain certain non-aqueous liquids, as well as small solid particulates such as salt-grains, pollens, smoke or dust (see 'condensation nuclei', below).

condensation nuclei

Minute airborne particles of dust or other solid material on which water vapour condenses into droplets; their presence is a necessary precondition for the formation of clouds. They occur naturally in the atmosphere in great abundance, although they have been added to as a result of human activity.

congestus (con)

From the Latin meaning 'to pile up' or 'accumulate', clouds of the species *congestus* grow higher than they are wide, with cauliflower-shaped tops (see Cumulus congestus ($C_L 2$)).

cumulogenitus (cugen)

A compound of the Latin words *cumulus* ('heap') and *genesis* ('origin') cumulogenitus clouds, such as Stratocumulus cumlulogenitus ($C_L 4$), are formed from the spread or decay of cumulus clouds.

Cumulonimbus (Cb)

A compound of the Latin words *cumulus* ('heap') and *nimbus* ('rain cloud'), convective Cumulonimbus clouds ($C_L 3$ and $C_L 9$) can grow to immense heights, producing lightning, hail, and stormy conditions at ground level.

Cumulus (Cu)

From the Latin word for 'heap' or 'pile', convective Cumulus clouds are generally detached, dense clouds, the upper parts of which appear brilliant white in sunshine.

duplicatus (du)

From the Latin for 'doubled' or 'repeated', clouds of the variety *duplicatus* persist in more than one layer.

fibratus (fib)

From the Latin for 'fibrous' or 'filamented', clouds of the species *fibratus* are nearly straight, with no hooks (as distinct from clouds of the species *uncinus*).

flammagenitus (flgen)

From the Latin for 'fire-made', the term denotes any cloud that forms over a direct source of heat, such as a wildfire or erupting volcano. Though previously termed 'pyrocumulus' clouds, the new term *flammagenitus* can be applied to clouds other than Cumulus.

floccus (flo)
From the Latin for 'tuft of wool', clouds of the species *floccus* are small and tufty, often with ragged lower parts.

fluctus (flu)
From the Latin for 'wave' or 'billow', this supplementary feature appears as distinctive wave-like crests above a variety of high- and mid-level clouds. *Fluctus* were formerly known as 'Kelvin–Helmholtz waves', after the 19th-century physicists who pioneered the study of turbulent flow.

flumen (flm)
From the Latin for 'flowing', *flumen* are bands of low accessory clouds associated with severe convective Cumulonimbus storm clouds, arranged parallel to the low-level winds and moving into or towards the storm cell.

fractus (fra)
From the Latin for 'broken' or 'fractured', clouds of the species *fractus* are ragged, sometimes patchy in appearance.

homogenitus (hogen)
From the Latin for 'man-made', the term applies to any clouds that arise from human activity, from aircraft contrails (Cirrus homogenitus) to the clouds that appear above industrial cooling towers (Cumulus homogenitus).

homomutatus (homut)
From the Latin *homo* ('man') and *mutatus* ('changed') the term refers specifically to persistent aircraft contrails that have spread out to cover much of the sky (see Cirrus homomutatus, p. 139).

humilis (hum)
From the Latin meaning 'low', or 'close to the ground', clouds of the *humilis* species are small, flattened, and generally wider than they are tall (see Cumulus humilis (C_L1), p. 24).

incus (inc)
From the Latin for 'anvil', an *incus* is an icy canopy that forms at the summit of Cumulonimbus capillatus clouds (see C_L9, p. 48; also p. 114).

intortus (in)
From the Latin for 'twisted' or 'entangled', clouds of this variety are irregularly curved or twisted in appearance.

irisation
Irisation refers to rainbow-like colours appearing at the edges of thin cloud as they pass across (or near to) the Sun or Moon. The term is derived from Iris, the Greek goddess of the rainbow (see p. 145).

lacunosus (la)
From the Latin for 'with gaps or holes', clouds of the variety *lacunosus* are reticulated like a net.

lenticularis (len)
From the Latin *lenticula* (meaning 'little lens' or 'lentil') clouds of the species *lenticularis* are almond or lens-shaped wave clouds, often formed by the movement of moisture-laden air over a high hill or mountain slope (see orographic, opposite).

mamma (mam)
From the Latin for 'udder' or 'breast', *mamma* are distinctive pouches hanging down from the under surfaces of Cumulonimbus clouds (see p. 115).

mediocris (med)
From the Latin for 'medium', clouds of the species *mediocris* are generally of equal width and depth, often with small bulges at the top (see Cumulus mediocris, C_L2).

murus (mur)

From the Latin for 'wall', *murus* denotes a large, localized lowering of cloud that develops beneath the surrounding base of a Cumulonimbus storm cloud. This supplementary feature is also known as a 'wall cloud'.

nacreous clouds

From the Latin word for 'mother-of-pearl', *nacreous* clouds, also known as polar stratospheric clouds, are icy clouds that form in the lower stratosphere, 15–30km (10–20 miles) above the Earth (see p. 128).

nebulosus (neb)

From the Latin for 'misty' or 'nebulous', clouds of the species *nebulosus* are thin, misty or veiled in appearance.

Nimbostratus (Ns)

A compound of the Latin words *nimbus* ('rain cloud') and *stratus* ('layer'), Nimbostratus (C$_M$2) is the name of the dense, grey blanket of cloud from which drizzle or persistent rain often falls.

noctilucent cloud (NLC)

From the Latin meaning 'night-shining', NLCs (also known as polar mesospheric clouds) are thin, icy clouds occurring high in the mesosphere, some 80km (50 miles) above the Earth (see p. 130).

opacus (op)

From the Latin for 'shadowy' or 'thick', clouds of the variety *opacus* completely mask light from the Sun or Moon.

orographic

From the Greek *oros* ('mountain') and *graphos* ('to write'), orographic clouds, such as wave clouds or banner clouds, are created or shaped by the presence of high mountains.

pannus (pan)

From the Latin for 'piece of cloth' or 'rag', *pannus* are accessory clouds in the form of ragged shreds that usually appear below rain clouds (see p. 111).

perlucidus (pe)

From the Latin meaning 'allowing light to pass through', clouds of the variety *perlucidus* allow some sun- or moonlight to be seen.

pileus (pil)

From the Latin for 'cap', *pileus* is a flattened, cap-shaped accessory cloud, occurring mostly with Cumulus and Cumulonimbus clouds (see p. 110).

praecipitatio (pra)

From the Latin for 'a fall', *praecipitatio* refers to precipitation of any kind (i.e. rain, snow or hail) that reaches all the way to the ground (as distinct from virga) (see p. 118).

pyrocumulus

A compound of the Greek word *pyro* ('fire'), and the Latin *cumulus* ('heap'), pyrocumulus is an earlier name for cumuliform clouds produced by combustion, whether natural (from volcanic eruptions or forest fires, now known as Cumulus flammagenitus), or human-made (from stubble-burning or industrial emissions, now known as Cumulus homogenitus) (see pp. 136–137).

radiatus (ra)

From the Latin for 'radiant' or 'radiated', clouds of the variety *radiatus* appear in parallel bands or rays which seem to converge.

silvagenitus (sigen)

From the Latin for 'forest-made', *silvagenitus* are special clouds created by the evaporation of moisture above large forested areas.

spissatus (spi)
From the Latin meaning 'thick' or 'condensed', clouds of the Cirrus species *spissatus* tend to be dense and grey in colour.

stratiformis (str)
A compound of the Latin words *stratus* ('layer') and *forma* ('form' or 'appearance'), stratiform clouds extend horizontally in a wide sheet or layer.

Stratocumulus (Sc)
A compound of the Latin words *stratus* ('layer') and *cumulus* ('heap'), Stratocumulus clouds appear as rounded masses or rolls of cloud which often seem to arrange themselves into parallel bands of cloud.

Stratus (St)
From the Latin word for 'layer' or 'sheet', Stratus clouds are low, sometimes indistinct layers of cloud, which seldom produce rain.

translucidus (tr)
From the Latin for 'translucent' or 'diaphanous', clouds of the variety *translucidus* show sun- or moonlight clearly through.

tuba (tub)
From the Latin for 'trumpet' or 'tube', *tuba* are funnel clouds which extend downwards from the base of a Cumulonimbus cloud, but rarely reach the ground (see p. 126).

uncinus (unc)
From the Latin for 'hooked', clouds of the species Cirrus uncinus (C_H1) are characteristically comma- or hook-shaped.

undulatus (un)
From the Latin meaning 'wavy', clouds of the variety *undulatus* feature parallel waves or undulations.

velum (vel)
From the Latin for 'ship's sail' or 'tent flap', *velum* is an accessory cloud of great horizontal extent, in the form of a veil enclosing the upper part of one or more cumuliform clouds (see p. 112).

vertebratus (ve)
From the Latin meaning 'in the form of vertebrae', clouds of this variety (typically Cirrus clouds) appear like ribs or fishbones in the sky.

virga (vir)
From the Latin for 'rod', *virga* ('fallstreaks') are streaks of precipitation (usually rain or snow) which fail to reach the ground (see p. 116).

volutus (vol)
From the Latin for 'rolled', *volutus* is a detached, tube-shaped cloud mass, that often appears to roll slowly about a horizontal axis cloud. This newly recognized species, added to the cloud classification in 2017, appears mostly with Stratocumulus and occasionally Altocumulus clouds.

Further Reading

Day, John A., *The Book of Clouds* (New York, 2006)

Dessler, Andrew E., *The Science and Politics of Global Climate Change: A Guide to the Debate*, 3rd edn (Cambridge, 2019)

Dunlop, Storm, *The Weather Identification Handbook* (Guilford, Conn., 2004)
Clouds: All You Need to Know in One Concise Manual (Yeovil, 2019)

Hamblyn, Richard, *The Invention of Clouds: How an Amateur Meteorologist Forged the Language of the Skies* (London, 2001)
Extraordinary Clouds (Newton Abbot, 2009)
Clouds: Nature and Culture (London, 2017)

Howard, Luke, *On the Modifications of Clouds, &c.* (London, 1803; 1865)
The Climate of London, 2 vols (London, 1818; 1820)

Kington, J. A., 'A Century of Cloud Classification', *Weather* 24 (1969): 84–89

Met Office, *Cloud types for observers: reading the sky* (Exeter, 2006)

Pretor-Pinney, Gavin, *The Cloudspotter's Guide* (London, 2006)
A Pig with Six Legs, and Other Clouds that Look Like Things (London, 2007)
A Cloud A Day (London, 2019)

Rubin, Louis D., and Duncan, Jim, *The Weather Wizard's Cloud Book: How You Can Forecast the Weather Accurately and Easily by Reading the Clouds* (New York, 1989)

Scorer, Richard, *Clouds of the World: A Complete Colour Encyclopedia* (Newton Abbot, 1972)
and Arjen Verkaik, *Spacious Skies* (Newton Abbot, 1989)

Stephens, Graeme L., 'The Useful Pursuit of Shadows', *American Scientist* 91 (2003): 442–49

Thornes, John E., *John Constable's Skies: A Fusion of Art and Science* (Birmingham, 1999)

Völter, Helmut (ed.), *Wolkenstudien | Cloud Studies | Études Des Nuages* (Leipzig, 2012)

World Meteorological Organization, *International Cloud Atlas*, 3 vols (Geneva, 1975; 1987; 2017)

Useful Links

https://cloudatlas.wmo.int/en/home.html
Homepage of the most recent (2017) edition of the World Meteorological Organization's *International Cloud Atlas*, the official guide to the classification of clouds.

www.cloudappreciationsociety.org
An online community for cloud lovers everywhere, founded by Gavin Pretor-Pinney, author of *The Cloudspotter's Guide* (2006) and *A Cloud A Day* (2019).

www.metoffice.gov.uk/weather/learn-about/weather/types-of-weather/clouds
The (UK) Met Office's informative cloud page.

https://www.noaa.gov/jetstream
The U.S. National Weather Service's online meteorology school.

https://www.rmets.org/
Homepage of the Royal Meteorological Society, founded 1850.

www.ametsoc.org
Homepage of the American Meteorological Society, founded 1919.

Index

Richard Hamblyn's books include *The Invention of Clouds*, which won the *Los Angeles Times* Book Prize; *Terra: Tales of the Earth*, a study of natural disasters; and *The Art of Science*, an anthology of readable science writing from the Babylonians to the Higgs boson. He has also written a trio of books in association with the Met Office: *Extraordinary Clouds*; *Extraordinary Weather*, and *The Met Office Pocket Cloud Book*. He teaches in the department of English, Theatre and Creative Writing at Birkbeck, University of London.

Picture Credits

Front cover © Unsplash/tcwillmott; 2 © Unsplash/jdiegoph; 4 © Unsplash/sendi_r_gibran; 6 © Unsplash/billy_huy; 8 © S Mallon; 11 ©V&A Images, Victoria & Albert Museum; 13, 15 © The British Library; 14 © Royal Meteorological Society; 20–1 © K Lew; 22, 23, 25, 38, 44, 76, 77, 84, 88, 89, 91, 97, 112, 138, 142, 144 © SD Burt; 26 © S Jebson; 27 © N Goodban; 28, 47, 58, 85 © M Clark; 29, 30, 31, 49, 51, 67, 102 (bottom), 107, 117, 118, 119, 126, 148, 149 © R Coulam; 32, 34, 35 (both), 40, 61, 63, 79, 80 (bottom), 82, 95, 102 (top), 104 (bottom) © RK Pilsbury; 33, 52, 55, 59, 70, 73, 75, 81, 87, 94, 96, 98, 99, 101 © CS Broomfield; 36 (bottom) © WG Pendleton; 36 (top), 37, 38 (bottom), 45, 54, 71, 110, 114, 139 © JFP Galvin; 39, 120, 122 (both), 124, 133, 134 (both) © Wikimedia Commons; 41 © A Bushell; 43, 46, 62, 78, 80 (top), 106, 116, 125, 150, 152, 153, 159 © Crown Copyright 2021, The Met Office; 48 (bottom) © JW Warwicker; 48 (top), 115, 128 © PJB Nye; 53 © R Ham; 56, 72 © KE Woodley; 57, 103 © WS Pike; 60, 66, 68 © C Irving; 64, 69, 93, 104 (top), 111 © J Corey; 65 © A Gilkes; 74 © SG Cornford; 86, 146 © A Best; 83 © JM Pottie; 90 © M Kidds; 92, 140 © J Walton; 100 © RW Mason; 105 © RD Whyman; 108–109 © S Jastrzebski; 113 © Steve Willington; 126 © P McKillop; 130 © M. Yrjölä; 132 © MJO Dutton; 136 © GJ Jenkins; 137 © CG Holmes; 138 © R Hamblyn; 141 © R Stagg; 143 © R Selby; 147 © TR Kenniston; 151 © DAR Simmons; 154–155 © D Freund; 157 © NASA/GSFC/Bob Cahalan; 161 © NASA/GSFC/Jaques Descloitres, MODIS Land Rapid Response Team; 163 © T Eklund. 165 © Unsplash/tijsvl; Back cover © Unsplash/alexmachado.

A DAVID AND CHARLES BOOK
© David and Charles, Ltd 2021

David and Charles is an imprint of David and Charles, Ltd
Suite A, Tourism House, Pynes Hill, Exeter, EX2 5WS

Text © Richard Hamblyn
Photographs © the copyright holders (see opposite)

First published in the UK and USA in 2008
This updated edition first published in 2021

Richard Hamblyn has asserted his right to be identified as author of this work in accordance with the Copyright, Designs and Patents Act, 1988.

A catalogue record for this book is available from the British Library.

ISBN-13: 9781446308905 paperback
ISBN-13: 9781446381083 EPUB

Printed in Slovenia by GPS Group for:
David and Charles, Ltd
Suite A, Tourism House, Pynes Hill, Exeter, EX2 5WS

10 9 8 7 6 5 4

Publishing Director: Ame Verso
Commissioning Editor: Neil Baber
Managing Editor: Jeni Chown
Editors: Katie Hardwicke and Sarah Tempest
Designers: Alistair Barnes and Blanche Williams
Pre-press Designer: Ali Stark
Production Manager: Beverley Richardson

David and Charles publishes high-quality books on a wide range of subjects. For more information visit www.davidandcharles.com.

Layout of the digital edition of this book may vary depending on reader hardware and display settings.